Population and Community Biology

MULTIVARIATE ANALYSIS OF ECOLOGICAL COMMUNITIES

Population and Community Biology

Series Editors

M. B. Usher
Reader, University of York, UK

M. L. Rosenzweig
Professor, Department of Ecology and Evolutionary Biology, University of Arizona, USA

The study of both populations and communities is central to the science of ecology. This series of books will explore many facets of population biology and the processes that determine the structure and dynamics of communities. Although individual authors and editors have freedom to develop their subjects in their own way, these books will all be scientifically rigorous and often utilize a quantitative approach to analysing population and community phenomena.

MULTIVARIATE ANALYSIS OF ECOLOGICAL COMMUNITIES

P. G. N. Digby

Department of Statistics
Rothamsted Experimental Station
Harpenden

and

R. A. Kempton

Scottish Agricultural Statistics Service
(formerly Agricultural and Food Research Council Unit of Statistics)
University of Edinburgh

Springer-Science+Business Media, B.V.

Originally published by Chapman & Hall in *1987*
Softcover reprint of the hardcover 1st edition *1987*

ISBN 978-94-010-7909-9 ISBN 978-94-009-3133-6 (eBook)
DOI 10.1007/978-94-009-3133-6

British Library Cataloguing in Publication Data

Digby, P. G. N.
Multivariate analysis of ecological communities.—(Population and community biology)
1. Biotic communities 2. Multivariate analysis
I. Title II. Kempton, R. A. III. Series
574.5'247'01519535 QH541.15.M34

Library of Congress Cataloging in Publication Data

Digby, P.G.N.
Multivariate analysis of ecological communities.

(Population and community biology)
Bibliography: p.
Includes index.
1. Biotic communities—Statistical methods.
2. Ecology—Statistical methods. 3. Multivariate analysis. I. Kempton, R.A., 1946– . II. Title.
III. Series.
QH541.15.S72D54 1987 574.5'247 86-17583
ISBN 978-94-010-7909-9

Contents

Preface

The last ten years have seen an enormous increase in the development and application of multivariate methods in ecology; indeed the perceived importance of these methods for elucidating the complex interactions observed in community studies is shown by the number of recent books devoted to introducing the more common multivariate techniques to ecologists (Williams, 1976; Orloci, 1978; Whittaker, 1978a, b; Gauch, 1982; Legendre and Legendre, 1983; Pielou, 1984) and by the chapters added to new editions of more general texts on quantitative ecology (e.g. Greig-Smith, 1983; Kershaw and Looney, 1985).

Two reasons can be put forward to explain this development. The first is undoubtedly the increasing availability of cheap computing power which makes it feasible to analyse the large data matrices involved in community studies. The second, perhaps less widely appreciated, is the change in emphasis of theoretical work on multivariate analysis, away from the development of formal statistical models and associated distribution theory towards descriptive techniques for exploring pattern in data sets and providing succinct summaries and displays. This new approach, termed 'pattern analysis' by Williams (1976), has led to a range of statistical techniques which have been enthusiastically taken up by ecologists to replace the collection of *ad hoc* procedures developed over the years for analysing community data.

This book brings together for the first time many of the new techniques of multivariate analysis appropriate for ecological data. The techniques include the familiar, and some less familiar, methods for ordination and classification and also some special techniques; for example, methods for analysing asymmetric association matrices and for comparing several different ordinations. Two preliminary chapters introduce the different types and forms of ecological community data and methods for preliminary inspection of data using graphs and tables. Little attempt is made to review early methods of analysis, many of which should now be of only historical interest: a good review of these early developments is given by Greig-Smith (1983).

A matrix approach is adopted for the theoretical development and this forms a unifying thread which links apparently disparate methods of analysis. It is expected that readers will have some familiarity with matrix algebra, but the Appendix contains the basic mathematical definitions and properties underlying the methods covered, and is included for reference and as a concise refresher course. Suitable elementary primers in matrix algebra are given in

Williams (1976) and Pielou (1984), although both of these texts subsequently cover a very limited range of multivariate techniques.

A particular strength of this book lies in the numerous illustrative examples. Several data sets are subject to successive reànalysis throughout the book, allowing comparison to be made of the results of different methods of analysis. Detailed case studies are also discussed at the ends of chapters, allowing a more considered interpretation of the results of multivariate analysis and leading in some cases to more formal modelling and analysis. Readers who are used to seeing the same familiar examples repeated in many recent textbooks will be pleased to know that, while the data sets here are in most cases drawn from previously published sources, the analyses presented are nearly all entirely new to this book. Most of these analyses were carried out using the Genstat statistical computer package and a final chapter includes details of the relevant multivariate directives and brief examples of the input instructions for the program.

The stimulation for this book originated from a Workshop on Multivariate Methods for Ecologists organized jointly by the British Ecological Society and Biometric Society at Edinburgh in September 1981. We are grateful to the participants who contributed ideas and examples to that Workshop and particularly to Ron Smith who was responsible for much of the organization and contributed to our initial thoughts about the form and structure for the book. We are also grateful to John Gower and Michael Usher who made many helpful criticisms of an early draft. We also thank especially Sue Land for her careful and patient preparation of the typescript. Our debt to John Gower, who has been a stimulating Rothamsted colleague for both of us, is particularly great and goes far beyond the many references to his pioneering work cited in this book.

P. G. N. Digby and R. A. Kempton
Rothamsted and Cambridge
December 1985

1 Ecological data

The development of statistical methods in field biology has, until recently, been dominated by the requirements of agricultural scientists working with highly controlled systems of experiments. Here the experimental factors (e.g. plant variety or nutrient level or site) generally cover only a limited range, while the variation in uncontrolled factors is made as small as possible, for example, by spraying to control disease. In consequence, the range of measured responses and the unexplained error in those responses are also fairly small. This has led to the development of a large statistical theory in which the pattern of response, possibly after transformation, is described by the effects of additive factors and a residual response whose distribution approximates to some known mathematical form (e.g. a normal distribution).

In contrast, data from ecological communities rarely conform to these model assumptions, even when they derive from formal experiments. Consider the typical example of Table 1.1, which we will be using on several occasions to illustrate the methods developed in this book. These data show the relative yields of meadow species recorded between 1973 and 1975 in plots from the famous Park Grass experiment laid out by John Bennett Lawes at Rothamsted Experimental Station in 1856 (Williams, 1978). Here of course plant species span a far wider range of genetic diversity than that usually found in crop trials and, after 130 years of continuous fertilizer treatment, the plots span a far wider range of soil nutrient status. In consequence the range of individual yields is far larger than normally observed in agricultural experiments; in particular, more than 50% of the cell entries in Table 1.1 are represented by zero yields, even though a number of rare species from the site have been completely omitted. Furthermore, the interaction between species and treatments, and among species within the same plot, is very complex. For these data, it is of interest to identify similarities in overall species response among the different treatments as well as associations among the different species themselves, but the standard procedures of statistical estimation and hypothesis testing are clearly inappropriate.

This book is concerned with methods for investigating the complex patterns which arise in ecological data of this type. However, before we look in detail at the different methods, we will first consider the commoner types and forms of such data and how one form may be constructed from another.

Table 1.1 Relative abundance of plant species (% total dry matter yield) under different fertilizer treatments on Park Grass plots 1973–75 (Williams, 1978). The 14 main plots are split for different liming treatments a, b, c and d.
t indicates abundance <0.05%,
. species not recorded.

	Plot number and fertilizer treatment																	
	3 Unmanured		8 PNaMg		7 PKNaMg		17 $N_1^{\dagger}$		16 $N_1^{\dagger}$PKNaMg		14 $N_2^{\dagger}$PKNaMg		1 N_1		18 N_2KNaMg		13 Organic	
Species	d	a	d	a	d	a	d	a	d	a	d	a	d	c	d	c	d	c
Agrostis tenuis	15.5	2.3	4.6	4.0	29.0	0.4	6.1	1.9	1.0	.	.	.	84.4	19.6	82.7	51.8	32.3	18.8
Alopecurus pratensis	2.5	1.0	2.8	1.0	7.3	5.2	24.4	7.9	28.8	4.3	36.8	13.4	.	.	.	.	16.1	15.3
Anthoxanthum odoratum	7.2	6.5	9.6	13.1	11.3	1.1	14.2	5.3	6.1	1.1	0.3	*t*	11.1	9.5	17.2	23.0	9.4	5.7
Arrhenatherum elatius	0.2	0.2	2.0	6.1	0.4	30.8	0.3	0.8	37.6	42.4	37.0	39.8	.	.	.	1.6	7.0	27.8
Briza media	1.0	2.0	0.1	1.6	.	.	0.2	3.9	.	.	.	.	.	.	.	.	.	.
Bromus mollis	.	.	.	.	*t*	0.9	.	.	.	0.1	1.3	2.4	.	.	.	*t*	.	.
Cynosurus cristatus	.	0.1	.	.	.	.	.	*t*	.	.	.	.	.	.	.	.	.	.
Dactylis glomerata	2.2	2.2	1.9	1.5	5.0	3.0	5.0	11.0	7.8	3.1	2.7	1.4	.	.	.	0.5	2.2	2.7
Deschampsia caespitosa	.	.	.	.	.	.	.	.	.	.	.	.	.	.	.	.	.	.
Festuca rubra	33.2	13.6	15.6	11.7	14.7	0.3	7.8	13.4	1.0	3.4	.	2.7	2.9	48.7	0.1	14.1	3.2	0.7
Festuca pratensis	.	.	0.3	3.6	.	0.5	0.3	.	.	0.2	.	2.0	.	.	.	.	.	.
Helictotrichon pubescens	0.3	8.5	0.3	12.0	0.1	0.3	0.4	16.7	0.6	1.5	.	4.2	.	.	.	.	.	.
Holcus lanatus	1.5	4.3	16.7	5.6	6.5	2.5	2.7	3.8	3.0	0.4	0.4	.	.	1.3	*t*	0.8	20.3	13.8
Poa pratensis	0.2	0.9	0.2	0.9	1.3	0.6	.	*t*	0.1	*t*	0.4	.	.	2.5	.	1.5	.	0.1
Poa trivialis	.	0.4	0.4	1.2	0.2	1.1	0.2	0.1	0.8	3.1	4.6	5.6	.	0.1	.	.	0.3	1.3

Lolium perenne	.	.	1.0	.	.	0.1	2.0	7.4	0.2	.	.	0.2	.	.	.	.	0.2	.
Trisetum flavescens	0.1	1.0	0.1	1.4	.	0.8	*t*	0.2	*t*	*t*	.	0.2	.	.	.	.	.	*t*
Lathyrus pratensis	0.5	1.7	0.3	0.1	5.4	14.0	0.1	.	1.8	4.9	1.5	8.4	.	0.5	.	*t*	0.2	5.4
Lotus corniculatus	2.5	3.0	1.1	2.0	0.1	.	.	1.0	.	.	.	.	.	.	.	.	.	.
Trifolium pratense	3.3	7.2	7.2	7.6	3.9	7.7	0.1	2.0	0.2	2.1	.	0.9	.	1.2	.	0.2	1.9	3.7
Trifolium repens	0.2	0.3	0.8	0.4	1.2	0.5	.	0.1	*t*	0.6	.	.	.	.	.	.	.	.
Achillea millefolium	1.3	1.3	0.9	1.6	0.7	*t*	0.4	1.1	0.1	0.1	.	.	.	0.2	.	0.3	1.2	0.3
Anthriscus sylvestris	.	.	.	.	.	0.6	.	.	.	3.1	8.3	4.7	.	.	.	.	.	0.3
Ajuga reptans	0.1	.	.	.	.	.	0.1	.	.	.	.	.	.	.	.	.	.	.
Carex caryophyllea	.	2.7	.	*t*	.	.	0.1	.	.	.	.	.	.	.	.	.	.	.
Centaurea nigra	1.5	1.2	2.3	1.7	2.5	.	1.5	1.2	.	.	.	.	.	1.5	.	0.4	0.2	0.1
Cerastium holosteoides	.	0.2	0.2	0.8	.	0.3	0.1	0.4	.	.	.	0.1	.	2.4	.	0.9	.	*t*
Conopodium majus	0.3	0.2	0.5	0.2	1.1	0.1	*t*	*t*	0.1	0.5	.	*t*	.	.	.	0.1	0.9	0.9
Galium verum	0.2	.	.	*t*	.	.	.	.	.	.	.	.	.	.	.	.	.	.
Heracleum sphondylium	.	.	.	.	0.3	7.2	0.1	0.7	0.1	18.1	0.1	1.5	.	.	.	0.7	1.1	0.1
Hypochaeris radicata	0.1	.	.	.	.	.	.	.	.	.	.	.	.	0.9	.	.	.	.
Knautia arvensis	0.4	0.3	1.0	0.4	.	.	.	.	.	.	.	.	.	.	.	.	.	.
Leontodon hispidus	10.0	12.2	8.3	6.2	.	.	4.4	5.4	.	.	.	.	.	0.1	.	*t*	.	.
Linum catharticum	.	*t*	.	.	.	.	.	.	.	.	.	.	.	0.2	.	.	.	.
Luzula campestris	1.4	2.2	2.9	2.2	1.2	.	0.3	0.7	.	.	.	.	1.0	1.1	.	*t*	0.1	*t*
Pimpinella saxifraga	0.3	0.4	*t*	0.3	.	.	*t*	0.9	.	.	.	.	.	0.7	.	.	.	.
Plantago lanceolata	6.1	10.3	16.2	9.5	6.7	3.3	23.9	12.4	2.9	1.6	.	*t*	.	4.8	.	0.2	1.8	0.9
Potentilla reptans	.	.	.	.	.	.	.	.	.	.	.	.	0.6	0.1	.	.	.	.
Poterium sanguisorba	6.9	11.7	.	.	.	.	.	0.4	.	.	.	.	.	0.3	.	.	.	.
Ranunculus acris	0.7	1.6	1.9	2.5	0.2	7.7	3.7	0.3	5.3	3.2	0.7	4.0	.	1.0	.	.	0.7	1.8
Rumex acetosa	.	0.1	0.1	0.3	0.2	1.9	0.9	.	1.0	1.4	1.5	0.8	.	2.7	.	2.5	.	.
Taraxacum officinale	0.1	0.3	0.6	0.4	0.4	8.8	0.9	0.4	1.4	4.8	4.5	7.9	.	0.7	.	0.9	0.7	0.2
Tragopogon pratensis	.	0.1	.	.	.	.	.	0.3	.	.	.	.	.	.	.	0.4	.	.
Veronica chamaedrys	.	*t*	.	*t*	.	.	.	.	.	.	.	.	.	.	.	.	.	.
Plot yield (t/ha)	0.8	1.6	2.8	2.3	3.2	4.9	2.1	2.4	4.4	4.6	4.8	4.4	0.6	2.1	1.1	3.3	3.7	4.2
Soil pH	5.2	6.8	5.2	6.8	5.0	6.4	5.7	7.1	5.3	6.6	5.8	6.8	4.1	4.3	3.9	4.4	4.9	5.0

Table 1.1 (continued)

	Plot number and fertilizer treatment																			
	4 N_2P				10 N_2PNaMg				9 $N_2PKNaMg$				11/1 $N_3PKNaMg$				11/2 $N_3PKNaMgSi$			
Species	d	c	a	b	d	c	a	b	d	c	a	b	d	c	a	b	d	c	a	b
Agrostis tenuis	23.9	21.0	6.9	6.9	30.7	36.0	2.2	3.4	14.7	11.3	.	.	.	*t*	.	.	1.6	1.6	.	0.2
Alopecurus pratensis	.	.	5.7	7.0	*t*	0.1	7.6	6.9	0.1	0.8	15.3	8.1	.	9.1	29.6	17.0	.	7.4	28.7	15.2
Anthoxanthum odoratum	76.1	4.9	7.0	9.6	69.0	7.8	16.3	17.9	71.7	11.1	0.2	1.5	5.1	*t*	.	.	0.3	*t*	0.2	.
Arrhenatherum elatius	.	.	0.5	.	*t*	0.1	19.6	22.3	.	2.7	53.4	50.7	.	30.0	38.3	63.9	.	27.5	50.3	60.8
Briza media	.	.	.	.	.	.	.	.	.	.	.	.	.	.	.	.	.	.	.	.
Bromus mollis	.	.	.	.	.	.	.	.	.	0.1	*t*	.	.	.	.	.	.	.	.	.
Cynosurus cristatus	.	.	.	.	.	.	.	.	.	.	.	.	.	.	.	.	.	.	.	.
Dactylis glomerata	.	.	*t*	.	.	1.4	.	.	.	.	1.9	4.3	.	6.3	2.3	2.0	.	5.7	1.3	4.8
Deschampsia caespitosa	.	.	.	.	.	0.2	.	.	.	.	.	.	.	.	.	.	.	.	.	.
Festuca rubra	*t*	55.9	53.1	40.9	0.1	24.1	38.9	36.6	.	3.6	0.1	1.5	.	2.0	.	0.1	.	4.1	0.1	.
Festuca pratensis	.	.	.	.	.	.	.	.	.	.	.	.	.	.	.	.	.	.	.	.
Helictotrichon pubescens	.	0.2	0.8	11.7	.	.	.	0.7	.	.	0.7	1.5	.	.	.	.	.	.	.	.
Holcus lanatus	.	2.7	2.2	1.5	0.1	21.2	3.9	6.8	13.5	44.3	6.5	5.6	94.9	36.7	21.8	7.1	98.1	30.9	10.6	2.8
Poa pratensis	.	12.2	19.5	15.8	0.1	7.4	5.8	2.5	.	16.1	2.4	1.3	.	11.7	2.0	2.9	.	13.4	1.2	1.8
Poa trivialis	.	2.9	.	.	.	0.2	.	.	.	0.6	1.5	*t*	.	3.4	0.9	0.6	.	7.8	3.0	10.1
Lolium perenne	.	.	.	.	.	.	.	.	.	.	.	.	.	.	.	.	.	.	.	.
Trisetum flavescens	.	.	.	.	.	.	.	.	.	.	.	0.5	.	.	.	.	.	.	.	.
Lathyrus pratensis	.	.	.	.	.	.	0.2	.	.	3.5	11.0	15.6	.	.	0.1	*t*	.	.	0.2	*t*

Lotus corniculatus	.	.	.	.	.	.	.	.	.	.	.	.	.	.	.	.	.	.	.	.
Trifolium pratense	.	.	.	.	.	.	.	.	.	*t*	.	0.4	.	.	.	.	.	.	.	.
Trifolium repens	.	.	.	.	.	.	.	.	.	.	.	.	.	.	.	.	.	.	.	.
Achillea millefolium	.	.	0.3	0.5	.	0.3	0.2	0.2	.	.	*t*	.	.	.	.	.	.	.	0.6	2.0
Anthriscus sylvestris	.	.	.	.	.	.	.	*t*	.	0.6	0.5	2.8	.	0.3	0.5	0.5	.	0.8	.	.
Ajuga reptans	.	.	.	.	.	.	.	.	.	.	.	.	.	.	.	.	.	.	.	.
Carex caryophyllea	.	.	.	.	.	.	.	.	.	.	.	.	.	.	.	.	.	.	.	.
Centaurea nigra	.	.	.	.	.	.	.	.	.	.	.	.	.	.	.	.	.	.	.	.
Cerastium holosteoides	.	.	.	.	.	.	.	*t*	.	.	*t*	.	.	*t*	.	.	.	0.2	.	.
Conopodium majus	.	.	.	.	.	.	.	*t*	.	.	.	*t*	.	.	.	.	.	.	.	.
Galium verum	.	.	0.3	0.5	.	.	0.1	0.3	.	.	.	.	.	.	.	.	.	.	.	.
Heracleum sphondylium	.	.	.	.	.	.	.	.	.	2.7	4.0	3.3	.	.	0.9	1.1	.	0.3	1.4	1.7
Hypochaeris radicata	.	.	0.5	.	.	.	.	.	.	.	.	.	.	.	.	.	.	.	.	.
Knautia arvensis	.	.	.	.	.	.	.	.	.	.	.	.	.	.	.	.	.	.	.	.
Leontodon hispidus	.	.	.	.	.	.	.	.	.	.	.	.	.	.	.	.	.	.	.	.
Linum catharticum	.	.	.	.	.	.	.	.	.	.	.	.	.	.	.	.	.	.	.	.
Luzula campestris	.	.	.	.	.	.	.	.	.	.	.	.	.	.	.	.	.	.	.	.

Table 1.1 (continued)

	Plot number and fertilizer treatment																			
	4 N_2P				10 N_2PNaMg				9 $N_2PKNaMg$				11/1 $N_3PKNaMg$				11/2 $N_3PKNaMgSi$			
Species	d	c	a	b	d	c	a	b	d	c	a	b	d	c	a	b	d	c	a	b
Pimpinella saxifraga	.	.	*t*	0.4	.	.	.	.	.	.	.	.	.	.	.	.	.	.	.	.
Plantago lanceolata	.	*t*	0.3	1.2	.	.	1.9	0.4	.	.	.	.	.	.	.	.	.	.	.	.
Potentilla reptans	.	.	.	.	.	.	.	.	.	.	.	.	.	.	.	.	.	.	.	.
Poterium sanguisorba	.	.	*t*	.	.	.	.	.	.	.	.	.	.	.	.	.	.	.	.	.
Ranunculus acris	.	.	0.2	.	.	0.1	0.1	0.1	.	.	.	.	.	.	.	.	.	.	.	.
Rumex acetosa	.	*t*	2.2	4.0	*t*	.	0.9	0.4	.	0.6	0.5	0.4	.	*t*	2.6	2.9	.	0.2	1.8	0.3
Taraxacum officinale	.	0.2	0.3	0.1	.	1.0	2.3	1.5	.	2.1	1.9	2.4	.	0.4	1.0	1.9	.	0.1	0.6	0.2
Tragopogon pratensis	.	.	.	.	.	.	.	.	.	.	.	.	.	.	.	.	.	.	.	.
Veronica chamaedrys	.	.	.	.	.	.	.	.	.	.	.	.	.	.	.	.	.	.	.	.
Plot yield (t/ha)	2.6	3.8	2.8	2.9	3.1	4.6	3.7	3.4	4.5	7.4	6.6	6.3	4.7	8.0	6.7	6.0	7.0	8.6	7.3	7.8
Soil pH	3.9	4.0	5.8	5.9	3.9	4.2	5.5	5.8	3.9	4.2	5.0	5.6	3.7	4.4	4.3	4.4	3.8	4.2	5.1	5.5

Treatments applied every year are:

N(1, 2, 3)	Ammonium sulphate supplying 48, 96 or 144 kg N ha^{-1}.
N†(1, 2)	Sodium nitrate supplying 48 or 96 kg N ha^{-1}.
P	35 kg P ha^{-1} as powdered superphosphate.
K	225 kg K ha^{-1} as potassium sulphate.
Na	15 kg Na ha^{-1} as sodium sulphate.
Mg	11 kg Mg ha^{-1} as magnesium sulphate.
Si	450 kg ha^{-1} of water-soluble powdered sodium silicate.
Organic	35 t ha^{-1} farm yard manure and fish meal to supply 63 kg N ha^{-1} every fourth year.
Lime	d plots were unlimed, a plots received 2.2 t CaO ha^{-1} as ground lime every fourth year. Plots c and b originally received same treatment as d and a respectively, but both received additional lime under a new liming scheme some years prior to sampling.

1.1 TYPES OF DATA

One of the most common types of data recorded by ecologists, particularly animal ecologists, is a count of the number of times an event occurs. This may be the number of individuals of a particular species which occurs in a quadrat, or the number of times one species is found in the vicinity of another species. In the first case the count is a measure of abundance, while in the second it is a measure of association. Several other types of abundance measures are commonly used, particularly in botanical studies: those listed by Greig-Smith (1983, Ch. 1) include yield, biomass, density and cover (i.e. the proportion of ground covered by the perpendicular projection of all aerial parts). These abundance measures are usually expressed in terms of unit area and are then independent of the adopted sample size. In contrast, another commonly used measure is the frequency of occurrence, i.e. percentage of samples occupied, which is closely related to the sample size. Sample abundance may often be assessed more rapidly with little effective loss of precision by visual estimation using a crude abundance score, say on a 1–9 scale. The Braun-Blanquet scale attempts to provide a single combined estimate of plant cover and number of individual plants. Other scales are based on either percentage cover or number of individuals and often use logarithmic class intervals (Table 1.2).

As an alternative to recording actual abundance it may be quicker simply to rank the species according to abundance. Although little used, species ranks may also be a more robust measure of their relative importance. Nevertheless, the ranking must be based on some underlying measure of abundance, and different measures may produce very different rankings when species of different sizes or behaviour are included (Table 1.3). Finally, we might consider making the simplest record of the presence or absence of each species at each site.

The choice of type of data to be collected will depend on the relative costs of collection and the level of precision required in the analysis. For large-scale biogeographical studies, where samples are very heterogeneous, a record of presence or absence will often be sufficient; but for local comparison of stands of vegetation, say, a quantitative measure of abundance will usually provide a substantial gain in information. Again, the choice of quantitative measure will depend on the objective of the survey and technical feasibility: if, for example, interest centres on productivity, a measure of biomass will be most appropriate; likewise, assessment of density is often not practicable, or sensible, in vegetation surveys because of the impossibility of recognizing discrete individuals. Greig-Smith (1983, Ch. 10) provides further discussion of these practical considerations.

In community studies, the primary data will usually be species abundance but the ecologist may also collect other data – typically of morphometric, behavioural, or environmental type – to relate to the abundances. If, for example, a species is shown to occur in only a proportion of sites in the sample,

Table 1.2 Three scales used in botanical sampling: the cover-abundance scale of Braun-Blanquet uses approximately additive class intervals for categorizing % cover, while the H-S-D scale adopts logarithmic class intervals (see Trass and Malmer, 1973, and Westhoff and Maarel, 1973). Lowe (1984) similarly adopted a logarithmic scale in assessing insect abundance.

Scale	*Braun-Blanquet*	*Hult–Sernander–Du Rietz*	*Lowe*
r	one or few individuals		
+	occasional and less than 5% cover of total plot area		
1	abundant and with very low cover, or less abundant with higher cover, but always less than 5% of total plot area	<6% cover (1/16 plot area)	1 individual
2	very abundant and less than 5% cover, or 5–25% cover	6–12.5% cover	2–3 individuals
3	25–50% cover, irrespective of number of individuals	12.5–25% cover	4–7 individuals
4	50–75% cover	25–50% cover	8–15 individuals
5	75–100% cover	50–100% cover	16–31 individuals
6	—	—	32–63 individuals

it is of interest to ascertain whether these sites have any environmental associations. Similarly, if a number of species are found to be grouped according to their habitat distribution, this grouping may be related to similarities in species morphology or behaviour. These ancillary data may be of many different types. They may be quantitative, e.g. soil pH, gradient and orientation of a sloping site, or time from an environmental disturbance, but made on entirely different scales of measurement so that some form of standardization may be desirable before analysis (see Section 1.3). Alternatively, they may be qualitative or nominal attributes which fall into a number of discrete states having no clear-cut order, such as soil or habitat type, method of management, wing pattern or flower colour. A special case of a qualitative attribute is that with only two observable states, e.g. sex. These are called binary attributes and are conventionally coded 0 and 1. A more detailed and comprehensive description of the different types of attributes is found in Clifford and Stephenson (1975) and Gordon (1981).

Table 1.3 Ranking of mollusc species from Scottish loch prior to pollution (Pearson, 1975), based on (i) number of individuals, (ii) biomass. Note that species show a 200-fold range in their individual sizes.

Species	*Rank* (i)	*No. of individuals*	*Total biomass* (g)	*Rank* (ii)
Myrtea spinifera	1	119	16.28	2
Lepidopleurus asellus	2	61	1.20	6
Lucinoma borealis	3	58	15.99	3
Thyasira flexuosa	4	28	1.29	5
Nucula sulcata	5	22	0.46	10
Astarte elliptica	6	20	48.74	1
Midiolus spp.	7	10	1.68	4
Corbula gibba	8	10	0.60	8
Thracia spp.	9	9	0.50	9
Cardiacca spp.	10	4	0.65	7
Abra spp.	11	2	0.04	12
Montacuta ferruginosa	12	2	0.02	13
Cultellus pellucidus	13	1	0.09	11

Missing data require special consideration and may be of several different types. Most straightforwardly, a data value may not have been recorded for one unit through oversight, or the record may have been subsequently lost. Secondly, in large surveys some records may be more detailed than others; for example not all field-workers may be able to identify all individuals down to the species level. In a third case, a character may not be recorded because it is dependent on the presence of another character. For example, in a taxonomic classification, an insect may be classified as having or not having wings. Those insects with wings may be further characterized according to wing pattern, but this secondary character is inapplicable for those insects without wings and is often treated as missing. Species abundance is sometimes considered as a special case of a dependent variable, since it can only be assessed when the species appears in the sample. Absence from the sample does not necessarily imply absence from the community as the species may still be present at a frequency too low to appear in a small sample fraction. This is an important consideration when comparing communities using samples of very different sizes. Differences in species profiles may then arise simply because the larger sample picks up the rarer species.

1.2 FORMS OF DATA

By the *form* of an ecological data set we mean the way in which the data are arranged. In the most common situation the data are naturally arranged to form a two-way table or matrix of, say, species by sites (Table 1.1). Then the abundance of the ith species at the jth site, denoted by x_{ij}, is located in the cell in the ith row and jth column of the matrix **X**. The complete matrix **X** thus has n rows corresponding to the number of species, and p columns corresponding to the number of sites. Alternatively, where we have environmental data for a number of variables at various sites, we can construct a matrix **Z** where z_{ij} is the data value for the jth environmental variable measured at the ith site: here, by convention, the rows now correspond to the sites and the columns to the variables.

In these two examples, the rows and columns of the data matrix have been classified by different sets of objects: species and sites for the abundance data, and sites and variables for the environmental data. As an alternative, data may be in the form of a square matrix where the rows and columns are classified by the same set of objects. Typically such data are measures of association between pairs of objects, for example, the number of times two particular plant species occur together at a site; here the complete matrix contains information for all pairs of species.

This latter example is typical of the majority of such data sets, in that the matrix of associations is symmetric, i.e. the association between the ith and jth species is the same as the association between the jth and ith species: $x_{ij} = x_{ji}$. Occasionally, however, an association matrix is asymmetric, as for example when x_{ij} is the number of times vegetation type j follows vegetation type i in an ecological succession.

In ecology, association matrices are not often obtained directly from sampling observations. Nevertheless, they are particularly useful for certain types of data analysis and are therefore frequently constructed from the more common rectangular data matrix. For example, a species-by-sites matrix of abundance data may be used to construct the symmetric matrix of pairwise associations among the species: a simple example of this is shown in the construction of Table 1.4 from Table 1.1; however, this can be done in various ways, which are discussed in detail in Section 1.4.

Many ecological studies will result in a number of data matrices. For example, a species-by-sites matrix of abundance data may be supplemented by a set of environmental variables measured at the sites: a simple example of such variables would be the geographical location of the sites, expressed as eastings and northings from a National Grid.

When a rectangular matrix of abundance data is obtained in each of several years, we have a three-way table of data classified by years, sites and species. Methods are available for handling such three-way tables directly; however, it is more common to analyse the species-by-sites matrix for each year separately and then use the techniques discussed in Chapter 4 to compare and combine the separate results. Four- (and higher-) way data matrices may also occur; for example, when species can be characterized according to their sex, we might have a four-way table of abundances classified by year, site, species and sex.

Table 1.4 The number of joint occurrences of major species on Park Grass plots (Table 1.1). Values on the diagonal indicate the total number of plots in which each species occurs.

	Species																				
1	*Agrostis*	29																			
2	*Alopecurus*	22	30																		
3	*Anthoxanthum*	28	27	35																	
4	*Arrhenatherum*	21	28	26	29																
5	*Dactylis*	17	24	22	25	25															
6	*Festuca rubra*	26	26	31	26	22	32														
7	*Helictotrichon*	13	16	17	15	14	17	17													
8	*Holcus*	27	29	32	28	24	29	16	35												
9	*Poa pratensis*	22	26	26	26	22	26	15	29	29											
10	*Poa trivialis*	17	23	22	23	22	22	13	24	22	25										
11	*Lolium*	6	7	7	7	7	7	6	6	4	7	7									
12	*Lathyrus*	15	21	20	22	20	20	12	22	20	20	6	23								
13	*Trifolium pratense*	14	15	17	16	15	17	12	16	14	15	7	16	17							
14	*Anthriscus*	6	13	11	13	11	11	6	12	12	12	2	10	6	13						
15	*Heracleum*	11	18	16	19	18	16	9	18	16	18	6	17	12	11	19					
16	*Leontodon*	8	6	8	7	7	8	6	8	7	6	3	7	8	0	3	8				
17	*Plantago*	18	17	20	17	15	20	15	19	17	14	7	15	15	5	10	8	20			
18	*Poterium*	5	4	5	4	4	5	4	5	5	3	1	3	4	0	1	4	5	5		
19	*Ranunculus*	16	18	19	18	16	18	13	18	16	15	7	15	14	6	10	7	17	5	19	
20	*Taraxacum*	23	28	28	28	25	28	17	30	28	25	7	23	17	13	19	8	20	5	19	31
		1	2	3	4	5	6	7	8	9	10	11	12	13	14	15	16	17	18	19	20

1.3 STANDARDIZATION AND TRANSFORMATION OF DATA

Ecological data matrices often require some form of internal standardization so that any dominant effects which are not of primary interest may be removed before analysis (Noy-Meir, 1973). In some cases a standardization is implicit in the particular analysis, but more usually it is left to the decision of the ecologist.

The most common form of standardization in a species-by-sites matrix is for differences in the species or site means. These are usually of less interest than the interrelationships between species and sites and may be simply artefacts of a somewhat arbitrary choice of abundance measure or sampling method. For example, the numbers of different moth species caught in a light trap may tell us more about the different photosensitivities of the species than about their relative frequencies at the site. Similarly, the efficiency of the sampling method may vary from site to site, under the different environmental conditions: for example, benthic sampling devices such as grabs are known to vary in efficiency when used on different substrates. These size effects may be removed by centring the data values about their species or site means,

$$x_{ij} \longrightarrow x_{ij} - x_{i.} \qquad \text{or} \qquad x_{ij} \longrightarrow x_{ij} - x_{.j},$$

or by double centring for both species and sites,

$$x_{ij} \longrightarrow x_{ij} - x_{i.} - x_{.j} + x_{..}.$$

(Here $x_{i.} = \Sigma_j x_{ij}/p$, $x_{.j} = \Sigma_i x_{ij}/n$ and $x_{..} = \Sigma_i \Sigma_j x_{ij}/np$, define the means of the ith species, jth site, and the overall mean respectively.) Alternatively, if the changes in population abundance are assumed to be multiplicative rather than additive, the abundances may be expressed as proportions by standardizing by species or site totals,

$$x_{ij} \longrightarrow x_{ij}/px_{i.} \qquad \text{or} \qquad x_{ij} \longrightarrow x_{ij}/nx_{.j}$$

as in Table 1.1.

A further standardization often applied after centring ensures that all columns of the data matrix have equal variance,

$$x_{ij} \longrightarrow (x_{ij} - x_{.j})/(\Sigma_i (x_{ij} - x_{.j})^2/n)^{1/2}.$$

This may be particularly necessary when working with environmental or morphometric data which are on different, and often arbitrary, scales of measurement. However, it cannot be generally recommended and is rarely justified when working with abundance data. Consider, for example, two species, one of which shows considerable geographical variation in its abundance, while the other is more ubiquitous and occurs with similar numbers at all sites. If these two species show similar variation within sites (e.g. among quadrats or years) one would expect the first species to be a better indicator of differences between sites and would wish this to be taken into account in the analysis. However, standardization to constant variance over sites would effectively remove real differences in species discriminant ability.

Some form of standardization is often required for naturally occurring association data. For example, in Table 1.4, the direct use of the number of joint occurrences of species as an index of association may be considered inappropriate because of the differences in their overall frequency of occurrence. A measure of association which is less affected by overall abundance may be obtained by standardizing a_{ij} (the number of joint occurrences of species i with species j) by, for example, the average number of occurrences of the two species,

$$a_{ij} \longrightarrow 2a_{ij}/(a_{ii}+a_{jj}).$$

The standardized measure of association now varies from 0, for species which never occur together in the same plot, to 1 (or 100%), for species which always occur together, regardless of the actual number of plots in which they occur. For example, in Table 1.5 the similarity between *Leontodon* and *Poterium* is 62%, although they only occurred together in four of the plots. However, while standardization has undoubtedly produced some improvement, a measure of association based on species presence or absence conveys only limited information for this data set. Further analysis of the Park Grass data in subsequent chapters is therefore based on the relative abundances of the species on the plots.

One attraction of standardizing variates by their standard deviation or range, is that the data values then become scale-independent. However, this may be achieved (save for an additive constant) without recourse to internal information from the data matrix by transforming the original values to logarithms. The logarithmic transformation, $x_{ij} \longrightarrow \log_b(x_{ij}+c)$, has a predominant place in quantitative ecology: it has the effect of compressing the upper end of the measurement scale and thus reducing the importance of large values relative to smaller values in the data matrix. The base of the logarithms b is usually chosen to be 2, e = 2.718, or 10: conversion from logarithms of one base to another is achieved by a simple multiplying factor, e.g. $\log_2 x = (\log_2 10) \log_{10} x = 3.32 \log_{10} x$. The constant c is usually chosen to take a small positive value so as to reduce differences between small data values and, particularly, to cope with zero values. For abundance data in which x is a count or percentage frequency, it is usual to take the constant $c=1$ or $c=\frac{1}{2}$.

Healy and Taylor (1962) considered a general family of power transformations $x \longrightarrow x^{1-k/2}$ where $k \neq 2$ is chosen to stabilize the variances of the transformed variate values when plotted against their means. The special case $k=2$ is represented by a logarithmic transformation, while $k=1$ gives a square-root transformation and $k=4$ a reciprocal transformation. In ecological work, choice of a non-integral value of k will generally be out of place in view of the refinement of the subsequent analysis; indeed, alternatives to the logarithmic transformation are used only infrequently. Many scales for

Table 1.5 Association of species on Park Grass plots based on the standardized number of joint occurrences (Czekanowski measure).

	Species																				
1	*Agrostis*	100																			
2	*Alopecurus*	75	100																		
3	*Anthoxanthum*	88	83	100																	
4	*Arrhenatherum*	72	95	81	100																
5	*Dactylis*	63	87	73	93	100															
6	*Festuca rubra*	85	84	93	85	77	100														
7	*Helictotrichon*	57	68	65	65	67	69	100													
8	*Holcus*	84	89	91	88	80	87	62	100												
9	*Poa pratensis*	76	88	81	90	81	85	65	91	100											
10	*Poa trivialis*	63	84	73	85	88	77	62	80	81	100										
11	*Lolium*	33	38	33	39	44	36	50	29	22	44	100									
12	*Lathyrus*	58	79	69	85	83	73	60	76	77	83	40	100								
13	*Trifolium pratense*	61	64	65	70	71	69	71	62	61	71	58	80	100							
14	*Anthriscus*	29	60	46	62	58	49	40	50	57	63	20	56	40	100						
15	*Heracleum*	46	73	59	79	82	63	50	67	67	82	46	81	67	69	100					
16	*Leontodon*	43	32	37	38	42	40	48	37	38	36	40	45	64	0	22	100				
17	*Plantago*	73	68	73	69	67	77	81	69	69	62	52	70	81	30	51	57	100			
18	*Poterium*	29	23	25	24	27	27	36	25	29	20	17	21	36	0	8	62	40	100		
19	*Ranunculus*	67	73	70	75	73	71	72	67	67	68	54	71	78	38	53	52	87	42	100	
20	*Taraxacum*	77	92	85	93	89	89	71	91	93	89	37	85	71	59	76	41	78	28	76	100
		1	2	3	4	5	6	7	8	9	10	11	12	13	14	15	16	17	18	19	20

rapid assessment of abundance use logarithmic class intervals (Section 1.1) and further transformation is then unnecessary. When abundance is described by rank order, it may be appropriate to use a rankit transformation (Fisher and Yates, 1963, Table 20). This has the effect of extending the scale at either end of the range and thus increasing separation among species of low rank, and among species of high rank, relative to the middle ranking species.

1.4 CONSTRUCTING ASSOCIATION DATA

Many terms are used for describing the association between pairs of units: similarity, proximity, dissimilarity, distance, as well as association itself. In this book we shall use the terms similarity and distance, although the latter is a generic term and may refer to a squared distance.

Similarities usually lie in the range zero to one; increasing similarity implies increasing likeness, and all self-similarities should equal the maximum possible value, e.g. one. Distances decrease with increasing likeness, they are usually non-negative and self-distances are zero. Both similarities and distances are symmetric, i.e. the distance between the *i*th and *j*th units is the same irrespective of whether it is measured from the *i*th unit or from the *j*th unit. Methods such as cluster analysis (Chapter 5) operate most naturally with similarity data, whereas some ordination methods, such as principal coordinates analysis (Chapter 3), are traditionally described in terms of distances. With some data it is usual to construct similarities, while other data lend themselves to the construction of distances. However, it matters little which form of association is chosen, since similarities can readily be transformed to distances, and vice versa. For example, similarities s_{ij} in the range zero to one can be transformed to distances d_{ij} by either of the methods

$$d_{ij} = 1 - s_{ij},$$
$$d_{ij} = -\log s_{ij},$$

although the former is more common.

This section describes how an $n \times p$ matrix $\mathbf{X}$ of units by variables can be used to construct a symmetric $(n \times n)$ matrix of associations among the n rows of $\mathbf{X}$. Of course we can always exchange rows and columns of $\mathbf{X}$ and so construct the symmetric $(p \times p)$ matrix of associations among the p columns of $\mathbf{X}$. We shall consider the formulation of association measures for three types of ecological data discussed earlier: binary, e.g. presence/absence; qualitative but with more than two states, e.g. colour; and quantitative, e.g. yield. We also need to consider what to do when the data are a mixture of these types and how to cope with missing values.

1.4.1 Binary data

When all the data are binary, information on the degree of association between any pair of units i and j may be displayed as a 2×2 contingency table:

		Unit j present	absent	total
Unit i	present	a	b	$a+b$
	absent	c	d	$c+d$
	total	$a+c$	$b+d$	p

where the total number of variables p is partitioned into those in which both units are present (a), those in which only one is present (b or c), and those in which both are absent (d). (If, as in Section 1.3, a_{ij} represents the number of times unit i occurs with unit j, $a=a_{ij}$, $b=a_{ii}-a_{ij}$ and $c=a_{jj}-a_{ij}$). Many different coefficients of similarity based on the quantities a, b, c and d have been proposed and the more important ones are given in Table 1.6: more comprehensive lists are given by Sneath and Sokal (1973), Goodall (1973) and Gower (1985).

All coefficients in Table 1.6 are symmetric in b and c, i.e. equal account is taken of the two situations: (i) when the ith unit is present and the jth is absent, and (ii) when the ith unit is absent and the jth is present. Clearly this is a necessary and sufficient requirement for the similarity coefficient itself to be symmetric, i.e. the similarity between units i and j is the same as that between j and i.

In contrast, symmetry in a and d will depend on the interpretation that may be placed on the data. Consider, for example, a binary variable characterizing plant species according to sexual type as monoecious (having male and female flowers on the same plant, coded as 1) or dioecious (male and female flowers on different plants, coded as 0). We would normally wish two dioecious species to be treated like two that were monoecious, and so would wish to use a similarity coefficient that was symmetric in a and d. The simple matching coefficient (Table 1.6) is the most commonly used measure with this property, although slight variants have been used in which the number of mismatches (b and c) carry either double or half weight (see Sneath and Sokal, 1973). However, an alternative interpretation may be more appropriate when the binary variable refers to the presence or absence of a taxonomic character or of a species at a site: here we may consider two species to be similar only if they positively occur together at the same sites, and may wish negative matches (conjoint absences) to be ignored in calculating the similarity coefficient. The most commonly used measure of this type is Jaccard's coefficient (coefficient 2 in Table 1.6). Coefficient 3 (which was used in deriving Table 1.5) and coefficient 4 are variants of coefficient 2 which are unlikely to lead to markedly different results in practical applications. The values of the four coefficients are shown in Table 1.7 for an artificial species-by-sites data matrix. Species A and C never appear at the same site, resulting in a zero similarity using coefficients 2, 3 and 4; but their similarity is 0.3, the same as between B and C, using coefficient 1.

Table 1.6 Similarity coefficients measures for binary data. Quantities *a*, *b*, *c* and *d* are defined in the text.

1.	$\frac{\text{co-occurrences plus conjoint absences}}{\text{total number of units}}$	$\frac{a+d}{a+b+c+d}$	Simple matching coefficient
2.	$\frac{\text{co-occurrences}}{\text{total occurrences over two sites}}$	$\frac{a}{a+b+c}$	Jaccard (1901)
3.	$\frac{\text{co-occurrences}}{\text{arithmetic mean occurrence at two sites}}$	$\frac{2a}{2a+b+c}$	Czekanowski (1913) Sørensen (1948)
4.	$\frac{\text{co-occurrences}}{\text{geometric mean occurrence at two sites}}$	$\frac{a}{\sqrt{(a+b)(a+c)}}$	Ochiai (1957)
5.	$\frac{\text{co-occurrences}}{\text{expected co-occurrences when independently distributed}}$	$\frac{a(a+b+c+d)}{(a+b)(a+c)}$	Mozley (1936) Margalef (1958)
6.	log-series measure	$\frac{2a}{a(b+c)+2bc}$	Mountford (1962)

Table 1.7 Similarities between three species based on presence and absence at ten sites using four different coefficients (Table 1.6).

		Site 1	2	3	4	5	6	7	8	9	10	*Total*
Species (1 present, 0 absent)	A	1	1	0	1	1	0	0	0	0	0	4
	B	0	1	1	1	1	1	1	1	1	0	8
	C	0	0	0	0	0	1	0	0	1	1	3

Summary tables (present +, absent −)

	Species B +	Species B −
Species A +	3	1
Species A −	5	1

	Species C +	Species C −
Species A +	0	4
Species A −	3	3

	Species C +	Species C −
Species B +	2	6
Species B −	1	1

	Species similarity		
Coefficient	AB	AC	BC
1 Simple matching	0.4	0.3	0.3
2 Jaccard	0.33	0	0.22
3 Czekanowski	0.5	0	0.36
4 Ochiai	0.53	0	0.41

Coefficients 1 to 4 all lie between 0 and 1. Coefficient 5 lies between 0 and p/a, where p is the number of sites used in classifying the species. Thus the similarity coefficient for pairs of species which always occur together, will be larger when both of these species are rare and occur in very few samples than when both are common. In particular, two species which occur together in a single sample and not elsewhere will be given a very high similarity, although this occurrence may be due purely to chance.

When deriving similarities among sites, one disadvantage of using the simple presence or absence of species is that results may be very sensitive to the sizes of the samples taken from the populations: in particular, the similarity between samples drawn from the same population, using for example Jaccard's coefficient, may be close to zero when sample size is small and the inclusion of a species is largely due to chance, but will approach 1 as sample size increases (Goodall, 1973). Coefficient 6 in Table 1.6 was derived by Mountford (1962) to be constant when comparing samples of different sizes drawn from the same population in which species abundances follow a logarithmic series distribution. Mountford also showed empirically that the log-series similarity coefficient was more robust to differences in sample size than Jaccard's or Czekanowski's measure when comparing different populations. Despite this, the coefficient has been little used, possibly because it relies on the abundances having a specific distributional form, which will not generally be true. A preferred course of action, when using presence–absence data, is to base comparisons on samples large enough for size differences to have only small effects.

1.4.2 Qualitative data

Sneath and Sokal (1973) proposed replacing multistate qualitative variables such as colour (red, white, blue) by a number of binary pseudo-variables (red/not red, white/not white, blue/not blue) to allow the use of one of the similarity coefficients in Table 1.6. However, we cannot recommend this procedure as it seems to be unnecessarily artificial and to lead to spurious weighting of variables with differing numbers of states.

Qualitative data are better handled using an extension of the simple matching coefficient:

$$s = \frac{\text{number of matching characters}}{\text{total number of characters}}.$$

When zero represents absence of a character, it may be preferable to ignore zero matches in an analogous way to the Jaccard coefficient. Alternatively, the character may be treated as two variables, a binary variable indicating presence or absence of the character, and a secondary multistate variable which is only specified when the character is present: different methods of constructing similarities for such hierarchical characters are considered by Gower (1971a).

1.4.3 Quantitative data

With quantitative data it is usual to operate with dissimilarities rather than similarities; although in some instances it is easier to work with measures of squared distance, rather than distance itself (see Section 1.4.5). Table 1.8 gives the more commonly used distance measures.

The Euclidean measure was the earliest used: this is the usual squared distance between pairs of units when the quantitative variables are plotted in coordinate space. The denominator r_k used in Table 1.8 is a normalizer, which is often included so that variables measured on differing scales are

Table 1.8 Distance measures d_{ij} between units i and j, based on p quantitative variables x_k ($k=1, \ldots, p$). r_k and w_{kl}^{-} are normalizers. In all cases summation is for $k=1, \ldots, p$.

1. Euclidean distance	$\frac{1}{p}\sum (x_{ik}-x_{jk})^2/r_k^2$	
2. City block (Manhattan)	$\frac{1}{p}\sum \lvert x_{ik}-x_{jk}\rvert/r_k$	Cain and Harrison (1958)
3. Bray–Curtis	$\frac{1}{p}\frac{\sum \lvert x_{ik}-x_{jk}\rvert}{\sum_k (x_{ik}+x_{jk})}$	Odum (1950) Bray and Curtis (1957)
4. Canberra	$\sum \frac{\lvert x_{ik}-x_{jk}\rvert}{(x_{ik}+x_{jk})}$	Lance and Williams (1966)
5. Mahalanobis	$\sum_l \sum_k (x_{ik}-x_{jk})w_{kl}^{-}(x_{il}-x_{jl})$	
6. Correlation coefficient complement	$1-\frac{\sum(x_{ik}-x_{i.})(x_{jk}-x_{j.})}{\sqrt{\sum (x_{ik}-x_{i.})^2 \sum (x_{jk}-x_{j.})^2}}$	

commensurate (Section 1.3): when applied, r_k is usually the range of the kth variable (this ensures that the contribution of any single attribute lies between 0 and 1), or alternatively its standard deviation.

An alternative distance measure, the Manhattan or city-block metric, is calculated as the total absolute difference between the variables, taking each in turn. Once again each variable may be standardized by its range (Gower, 1971a), but alternative formulations have been proposed (e.g. measures 3 and 4 in Table 1.8).

When working with abundance data, the main difference between the unstandardized Euclidean and Manhattan metrics lies in the dominance given to the larger data values. Since the Euclidean metric is a function of squared differences it is more sensitive to large aberrant values: this effect may be reduced by taking a logarithmic transformation of the data before calculating distances. Table 1.9 compares the range-standardized versions of these two metrics with the Canberra and Bray–Curtis distance measures. As in Table 1.8, all measures are standardized to lie between 0 and 1 by dividing by the total number of species. The Bray–Curtis measure is particularly sensitive to large outlying values: thus sites A and B differ only in the abundance of species 3 and the other three measures show this pair of sites to be the shortest distance apart; however, the high value of 55 individuals for species 3 at site B has completely dominated the Bray–Curtis measure so that A and C now appear to have the shortest distance. In addition to its extreme sensitivity to outlying values, the Bray–Curtis measure has the disadvantage of not being a metric

Table 1.9 Construction and comparison of distance measures for a small artificial data set.

		Species						
		x_1	x_2	x_3	x_4	x_5		
Site	A	10	4	5	1	0		
	B	10	4	55	1	0		
	C	5	8	10	5	2		
Range		5	4	50	4	2	*Total*	
$\|x_B - x_C\|$		5	4	45	4	2	60	
$x_B + x_C$		15	12	65	6	2	100	
								Distance BC
Euclidean		1	1	0.81	1	1	4.81	4.81/5
Manhattan		1	1	0.90	1	1	4.90	4.90/5
Canberra		0.33	0.33	0.69	0.67	1	3.02	3.02/5
Bray–Curtis†		(0.25)	(0.20)	(2.25)	(0.20)	(0.10)		60/100

		Euclidean	*Manhattan*	*Canberra*	*Bray–Curtis*
Inter-site	d_{AB}	0.20	0.20	0.17	0.56
distances	d_{AC}	0.80	0.82	0.53	0.40
	d_{BC}	0.96	0.98	0.60	0.60

Euclidean and Manhattan metrics are standardized by species range.
† The Bray–Curtis measure is derived directly from the totals for $|x_B - x_C|$ and $x_B + x_C$, but the contributions from the individual species are given for comparison with the other measures.

measure (see Section 1.4.5): in consequence we cannot recommend its general use in ecology.

For the most part we shall be using the Euclidean and Manhattan metrics in this book. For the reasons given in Section 1.3 we do not generally advocate scaling by the individual species ranges or standard deviations: if distances are required to lie between 0 and 1, for example to allow transformation to similarities, this can be achieved by dividing by the maximum derived distance or the overall range of abundance within the species-by-sites table.

Table 1.8 also includes two less commonly used distance measures. The Mahalanobis distance (measure 5) may be useful when units come from a number of known groups. A variance/covariance matrix $\mathbf{W}_g$ $(p \times p)$ is first calculated for each group g, and a pooled within-group dispersion matrix, $\mathbf{W} = \Sigma \mathbf{W}_g$, constructed. The weighting element w_{kl}^{-} is the (k, l)th element of the inverse matrix $\mathbf{W}^{-1}$. In the particular case when the within-group correlations among the variables can be ignored, the Mahalanobis distance approximates to the Euclidean squared distance with inverse weights r_k^2 $(k = 1, \ldots, p)$ given by the within-group variances for the p variables. Because the Mahalanobis distance is invariant to changes in scale of any of the variables, it is especially useful for variables with widely differing scales of measurement: when available, it is in our view much more effective than scaling by species range. Strictly speaking, the use of a pooled within-groups dispersion matrix assumes that the within-group variability is homogeneous, but this assumption is often invalid: a transformation of the variables may help.

The correlation coefficient has had widespread use as a similarity coefficient in taxonomic studies, and its complement may be used as a distance measure (measure 6 in Table 1.8). Its popularity arises, in part, because it uses the variables adjusted for the mean value over all variables, for each species, and thus ignores differences in their overall sizes. Such adjustment is questionable, unless all the variables are on the same scale of measurement.

1.4.4 A general form of similarity coefficient

Gower (1971a) proposed a general form of similarity coefficient which contains many of those previously discussed as special cases and copes logically with the situation of missing values and mixed variable types. The basic idea is to define a similarity coefficient s_{ijk} between the ith and jth unit as given by the kth variable, and then to average s_{ijk} over all variables k to form an overall similarity measure s_{ij}.

In the simplest situation all variables are treated alike and an unweighted average is calculated; for example, the simple matching coefficient for binary or multistate qualitative variables can be constructed in this way by taking $s_{ijk}=1$ when the data values of the kth variable for the ith and jth unit, x_{ik}, x_{jk}, are equal, and $s_{ijk}=0$ otherwise.

More generally, some form of weighted average may be desirable. For example, if some variables are known to be good discriminators between groups of units, we might weight these more highly to emphasize separations among the groups. Alternatively, when calculating a similarity coefficient among stands of vegetation based on species abundance, we may wish to weight species according to size, to give an overall measure based both on cover and abundance. Even more generality is obtained by allowing the weights to depend on the data values themselves. If $w_{ijk}=w(x_{ik}, x_{jk})$, a function of x_{ik} and x_{jk}, we have

$$s_{ij}=\sum_{k=1}^{p} w_{ijk}s_{ijk}\Big/\sum_{k=1}^{p} w_{ijk}.$$

This allows missing values to be treated individually without omitting the whole variable, by setting $w_{ijk}=0$ whenever either data value x_{ik} or x_{jk} is missing. Furthermore, it also allows double-zero matches to be ignored by setting $w_{ijk}=0$ when $x_{ik}=x_{jk}=0$. Table 1.10 shows the choices of w_{ijk} which lead to the three most commonly used similarity coefficients for binary data (measures 1, 2 and 3 in Table 1.6).

Table 1.10 Scores s_{ijk} and weights w_{ijk} for constructing different similarity coefficients for units *i* and *j* with binary character *k*. SMC denotes the simple matching coefficient.

Character value			w_{ijk}		
x_{ik}	x_{jk}	s_{ijk}	SMC	Jaccard	Czekanowski
1	1	1	1	1	2
1	0	0	1	1	1
0	1	0	1	1	1
0	0	1	1	0	0

So far we have considered only the coefficients used with binary or qualitative variables. To include quantitative variables we first calculate inter-unit distances d_{ijk} for each variable k, as in Section 1.4.3, and then transform them to similarities s_{ijk}, using for example $s_{ijk}=1-d_{ijk}$: this transformation requires that $0 \leqslant d_{ijk} \leqslant 1$, which is achieved typically by an appropriate choice of normalizer r_k in Table 1.8. This approach was adopted in constructing the distance measures in Table 1.9, although the final conversion to a similarity measure was not made. Thus, if considered desirable, double-zero matches can be ignored in calculating similarities for quantitative data using Euclidean or Manhattan distances, in a way exactly analogous to binary data. In Table 1.9 this would involve dividing the summed distances between A and B by 4 rather than 5, so that the Euclidean and Manhattan metrics both become 0.25, representing similarities of 0.75.

1.4.5 Properties of measures of association

Earlier in this section we described several ways in which similarities or distances can be calculated from a sites-by-species table, or a units-by-variables data matrix. Obviously the choice of which method to use will depend on the type of data, and we have already given some suggestions to help this choice. When associations are being calculated as a first stage towards an ordination, i.e. representing the units as a set of points in space, there are two properties of measures of similarity, or distance, that are generally considered to be desirable. The first property is relevant for any ordination method; the second is particularly relevant for principal coordinates analysis, which we describe in Section 3.5.

Both properties, of 'metricity' and 'Euclideanarity', are best thought of in terms of distance rather than similarity. As noted earlier, distance is a generic term and often implies squared distance. To avoid any confusion we will use δ

to represent actual distance; so δ_{ij} is the actual distance between the ith and jth species, say. Sometimes it is clear how actual distance should relate to the coefficients of similarity or distance given in Tables 1.6 and 1.8. For example, Euclidean distance, averaged over p variables, can be given as

$$d_{ij} = \frac{1}{p} \sum_{k=1}^{p} (x_{ik} - x_{jk})^2 / r_k^2 .$$

This is actually a squared distance, so it is usual to take $\delta_{ij} = \sqrt{d_{ij}}$. The same is true of Mahalanobis distance; however, it is not obvious whether, for example, city-block distances are actual distances or squared distances. Likewise it is unclear whether the simple transformation from similarity values $d_{ij} = 1 - s_{ij}$ is to represent an actual distance or a squared distance.

The *metric* property relates directly to the triangle inequality. For any three actual distances δ_{12}, δ_{13} and δ_{23}, the triangle inequality is $\delta_{ij} \leqslant (\delta_{ik} + \delta_{jk})$ for any allocation of 1, 2 and 3 to i, j and k; i.e. it is possible to construct a triangle with sides δ_{12}, δ_{13} and δ_{23}. A distance measure is metric if it always generates actual distances that will satisfy the triangle inequality. For example, consider the two variables given below for three species.

Species	*V*1	*V*2
1	0	1
2	1	0
3	1	1

Using the Bray–Curtis distance (given in Table 1.8), the inter-species distances are: $d_{12} = 1/2$, $d_{13} = 1/6$ and $d_{23} = 1/6$. Setting $\delta_{ij} = d_{ij}$, $\delta_{12} > \delta_{13} + \delta_{23}$, and it is impossible to construct a triangle with sides (1/2, 1/6, 1/6); therefore the metric property is not satisfied.

The importance of this property lies in ordination: no suitable ordination of these three species can be obtained to satisfy their Bray–Curtis distances. Obviously, the Bray–Curtis distance measure does not always give non-metric distances, for example those obtained in Table 1.9 do satisfy the triangle inequality. However, for many measures of distance it can be shown that the triangle inequality will always be satisfied; for example if Euclidean distances are calculated, the distance measure $\delta_{ij} = \sqrt{d_{ij}}$ is always metric. This is fairly easy to see: if the data values are considered as giving the coordinates of points in space, the square root of Euclidean distance is simply the straight-line distance between pairs of points; since the points lie in a real space, it is always possible to construct triangles among them. However, (squared) Euclidean distance itself is not metric: the example above generates inter-species values $d_{12} = 2$, $d_{13} = 1/2$, and $d_{23} = 1/2$. Table 1.11 shows which measures always give distances satisfying the metric property. For similarity measures it is usual to

Table 1.11 Metric and Euclidean properties of measures of association (where known).

	Metric		*Euclidean*	
Similarity coefficients	$\delta_{ij}=(1-s_{ij})$	$\delta_{ij}=\sqrt{(1-s_{ij})}$	$\delta_{ij}=(1-s_{ij})$	$\delta_{ij}=\sqrt{(1-s_{ij})}$
Simple matching	yes	yes	no	yes
Jaccard	yes	yes	no	yes
Czekanowski	no	yes	no	yes
Ochiai	no	yes	no	yes
Mozley	?	?	?	?
Mountford	?	?	?	?
Distance measures	$\delta_{ij}=d_{ij}$	$\delta_{ij}=\sqrt{d_{ij}}$	$\delta_{ij}=d_{ij}$	$\delta_{ij}=\sqrt{d_{ij}}$
Euclidean	no	yes	no	yes
City-block	yes	yes	no*	?
Bray–Curtis	no	?	no	?
Canberra†	yes	yes	no	?
Mahalanobis	no	yes	no	yes
Correlation complement	no	yes	no	yes

* The city-block metric is Euclidean if the normalizer is the range.
† The properties for this measure assume the data values are all positive.

consider whether $\delta_{ij}=1-s_{ij}$, or $\delta_{ij}=\sqrt{(1-s_{ij})}$ are metric: Table 1.11 also gives this information for most of the similarity coefficients of Table 1.6.

The *Euclidean* property is less general than that of metricity and relates to a complete set of distances. A distance matrix is Euclidean if all the actual distances could arise as the straight-line distances among a set of points in a real space. A distance measure is Euclidean if it always gives rise to distance matrices that are Euclidean. Consider the example below of the distances among four points, for various values of x.

unit				
1	0			
2	2	0		
3	2	2	0	
4	x	x	x	0
	1	2	3	4

The first three units can be represented by points placed in a two-dimensional space at the three corners of an equilateral triangle of side 2. If $x<1$ the distances are not metric, because none of the triangles involving the fourth unit can be constructed. If $x=1$ the distances satisfy the metric property; however they are not Euclidean because the fourth unit cannot be located at one point to satisfy all the distances (it would need to be simultaneously at the mid-point of each side of the triangle). If $x = \frac{2}{3}\sqrt{3}$ ($\simeq 1.15$) the fourth point can be located at the centre of the triangle to satisfy the distances, so they are now Euclidean, as well as being metric. If x is larger than this, the fourth point can be located above (or below) the plane of the triangle, so that the four points are at the vertices of a tetrahedron in three dimensions. This raises an interesting point: if a set of distances among n units is Euclidean, at most $(n-1)$ dimensions are needed to represent them. This follows from the example above, where the fourth point introduces a third dimension (if $x > 1.15$); each extra point may introduce one more dimension, but no more than one.

Clearly, any distance measure that is Euclidean will also be metric, but the converse is not true. Table 1.11 shows which of the similarity coefficients and distance measures are Euclidean. Because distance is sometimes used in the sense of squared distance, the properties are also given for the square root of each measure. Note, however, that these properties break down if the data set contains missing values. A fuller list of measures, and more discussion of these properties, is given by Gower and Legendre (1986), from which the information of Table 1.11, has been abstracted.

2 Preliminary inspection of data

In this book we place great emphasis on the visual display of the results of multivariate analysis. However, before becoming involved in more complicated analyses, we here consider some simple graphical methods suitable for a preliminary inspection of the data. These methods will often provide an insight into the data, highlighting aberrant points which may be unduly influential and allowing a check on the underlying assumptions of a proposed analysis; equally they may suggest directions for further analysis. Tukey (1977) coined the term 'exploratory data analysis' to describe this preliminary inspection of data. The reader will find a large number of original and ingenious graphical methods for displaying multidimensional data in Tukey and Tukey (1981) and Tufte (1983), while Green (1979) contains interesting ecological examples.

2.1 DISPLAYING DATA VALUES

Displaying data, e.g. abundances of species over sites, as a table of raw numbers will rarely allow a quick visual assessment. Ehrenberg (1982) makes several useful suggestions for displaying tables of data: one is that values should be displayed to two effective digits, i.e. to the first two digits that vary appreciably in the data. For example, with the values 16.79, 18.32, 19.81, 22.75, the first digit only varies between 1 and 2, so that the two effective digits are the second and third: hence the figures would be displayed as 16.8, 18.3, 19.8, 22.8. With large amounts of data, a more severe rounding may be appropriate.

Tukey (1977) discusses the 're-expression' of data values before displaying them; for example, data which are counts are usually best logged. Changing the scale on which data are displayed can also be effective. The scale 0 to 19 is particularly useful as the eye easily distinguishes between those values greater than and those less than 10. For counts in the range $0 \leqslant x \leqslant 100$, one way of achieving this is to express each as $10 \log_{10}(x+1)$ to the nearest integer. Of course, these rounded values are only for display; the actual values should always be used for any analysis.

The pattern of values in a two-way table is often much clarified by reordering the rows and/or columns. This forms the basis of the Braun–Blanquet approach, which was one of the earliest methods for classification of stands of vegetation, and the general effectiveness of reordering will be

Table 2.1 Pecking relationship among 32 hens. A value of 1 in a cell indicates that the bird in the row is dominant over the bird in the column; ½ indicates no clear dominance.

(a) Birds unsorted

Bird	YG	bY	bB	Y	GR	B	nn	nR	Gb	G	RY	BG	RG	GG	Gn	nG	BB	nb	Yb	RB	YR	BY	bG	R	Yn	2Y	bb	YY	bR	nB	Rb	n
YG		1		1				1							1			1	1			½						1	1	1		
bY								1											1										1			
bB	1	1		1	1	1	1	1	1	1	1	1	1	1	1	1	1	1	1	1	1	1	1	1	1	1	1	1	1	1	1	1
Y		1						1									1		1						1			1				
GR	1	1		1		1		1	1				1				1	1	1			1	1		1			1	1	1		
B	1	1		1				1									1		1				1		1			1	1	1	1	
nn	1	1		1	1	1		1				1			1	1	1	1	1	1	1	1	1		1	½		1	1	1		
nR																									1			1	1			
Gb	1	1		1		1	1	1									1	1	1						1			1		1		1
G	1	1		1	1	1	1	1	1			1	1		1		1	1	1	1	1	1	1	1	1		1	1	1	1	1	1
RY	1	1		1	1	1	1	1	1	1		1	1	1	1		1	1	1	1	1	1	1	1	1		1	1	1	1	1	1
BG	1	1		1	1	1		1	1				1				1	1	1		1		1		1			1	1	1	1	
RG	1	1		1		1	1	1	1						1		1	1	1		1	1	1		1			1	1	1	1	
GG	1	1		1	1	1	1	1	1	1		1	1		1	1	1	1	1	1	1	1	1	1	1		1	1	1	1	1	1
Gn		1		1	1	1		1	1			1							1		1		1		1			1	1	1	1	
nG	1	1		1	1	1		1	1	1	1	1	1		1		1	1	1	1	1	1	1	1	1	1	1	1	1	1	1	1
BB	1	1						1							1							1						1	1			
nb		1		1		1		1							1		1		1				1		1			1	1			
Yb								1									1											1	1			
RB	1	1		1	1	1		1	1			1	1		1		1	1	1		1	1	1		1		1	1	1	1	1	1
YR	1	1		1	1	1		1	1								1	1	1				1		1			1	1	1		
BY	½	1		1		1		1	1			1			1			1	1		1		1		1			1		1		
bG	1	1		1				1	1								1		1						1			1		1		
R	1	1		1	1	1	1	1	1			1	1		1		1	1	1	1	1	1	1		1		1	1	1	1	1	1
Yn	1	1															1		1									1	1			
2Y	1	1		1	1	1	½	1	1	1	1	1	1	1	1		1	1	1	1	1	1	1	1	1		1	1	1	1	1	1
bb	1	1		1	1	1	1	1	1			1	1		1		1	1	1		1	1	1		1			1	1	1	1	
YY		1																											1			
bR				1					1													1	1									
nB		1		1				1									1	1	1						1			1	1			
Rb	1	1		1	1		1	1	1								1	1	1		1	1	1		1			1	1	1		
n	1	1		1	1	1	1	1				1	1		1		1	1	1		1	1	1		1		1	1	1	1	1	

(b) Birds sorted on pecking score

Bird	bB	2Y	nG	RY	GG	G	R	RB	n	bb	nn	RG	BG	Rb	GR	Gn	YR	BY	Gb	B	nb	bG	YG	nB	BB	Y	Yn	Yb	bR	bY	nR	YY	*Pecking score*	*Rank*
bB		1	1	1	1	1	1	1	1	1	1	1	1	1	1	1	1	1	1	1	1	1	1	1	1	1	1	1	1	1	1	1	31	1
2Y				1	1	1	1	1	1	1	½	1	1	1	1	1	1	1	1	1	1	1	1	1	1	1	1	1	1	1	1	1	28½	2
nG		1		1		1	1	1	1	1		1	1	1	1	1	1	1	1	1	1	1	1	1	1	1	1	1	1	1	1	1	28	3
RY					1	1	1	1	1	1	1	1	1	1	1	1	1	1	1	1	1	1	1	1	1	1	1	1	1	1	1	1	28	4
GG			1			1	1	1	1	1	1	1	1	1	1	1	1	1	1	1	1	1	1	1	1	1	1	1	1	1	1	1	28	5
G							1	1	1	1	1	1	1	1	1	1	1	1	1	1	1	1	1	1	1	1	1	1	1	1	1	1	26	6
R								1	1	1	1	1	1	1	1	1	1	1	1	1	1	1	1	1	1	1	1	1	1	1	1	1	25	7
RB									1	1		1	1	1	1	1	1	1	1	1	1	1	1	1	1	1	1	1	1	1	1	1	23	8
n										1	1	1	1	1	1	1	1	1		1	1	1	1	1	1	1	1	1	1	1	1	1	22	9
bb											1	1	1	1	1	1	1	1	1	1	1	1	1	1	1	1	1	1	1	1	1	1	22	10
nn		½	1					1					1		1	1	1	1		1	1	1	1	1	1	1	1	1	1	1	1	1	20½	11
RG											1			1		1	1	1	1	1	1	1	1	1	1	1	1	1	1	1	1	1	19	12
BG												1		1	1		1		1	1	1	1	1	1	1	1	1	1	1	1	1	1	18	13
Rb											1				1		1	1	1		1	1	1	1	1	1	1	1	1	1	1	1	17	14
GR												1						1	1	1	1	1	1	1	1	1	1	1	1	1	1	1	16	15
Gn													1	1	1		1		1	1		1		1		1	1	1	1	1	1	1	15	16
YR															1				1	1	1	1	1	1	1	1	1	1	1	1	1	1	15	17
BY													1			1	1		1	1	1	1	½	1		1	1	1		1	1	1	14½	18
Gb									1		1									1	1		1	1	1	1	1	1		1	1	1	13	19
B													1								1	1	1	1	1	1	1	1	1	1	1	1	12	20
nb																1				1		1			1	1	1	1	1	1	1	1	11	21
bG																			1				1	1	1	1	1	1		1	1	1	10	22
YG																1		½			1			1		1		1	1	1	1	1	9½	23
nB																					1				1	1	1	1	1	1	1	1	9	24
BB																1		1					1						1	1	1	1	7	25
Y																									1		1	1		1	1	1	6	26
Yn																							1		1			1	1	1		1	6	27
Yb																									1				1		1	1	4	28
bR																		1	1			1				1							4	29
bY																												1	1		1		3	30
nR																											1		1				3	31
YY																													1	1		1	2	32

demonstrated extensively in future chapters. The example in Table 2.1 shows the pecking relationships among a flock of 32 hens (Guhl, 1953). Here the birds initially appear in random order. A pecking score may be calculated for each bird to indicate the number of birds over which it is dominant and this forms a natural basis for reordering rows and columns. When this is done the relationship among the birds is seen to approximate to a linear hierarchy: deviations from that hierarchy are immediately visible, and are investigated further in Chapter 6.

Rather than displaying a table of rounded numbers, each value could be replaced by a shaded grid square with the density of shading increasing for larger values. Figure 2.1 gives an example of a table of percentage faunal

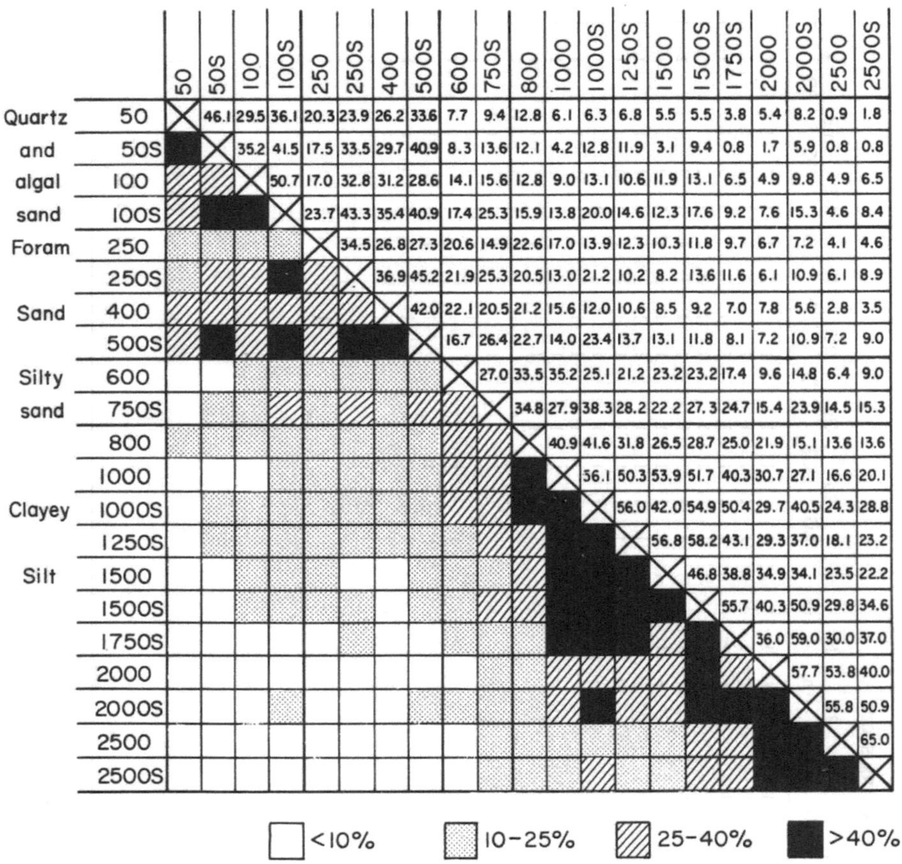

Fig. 2.1 Percentage faunal affinity of samples of nematodes taken at various depths from the sea bottom off North Carolina. The upper triangular matrix gives the calculated similarities among pairs of samples, the lower triangular matrix gives their shaded representation. (After Tietjen, 1971, Fig. 3.)

similarities among substrate samples at different ocean depths. As the table is symmetric about the leading diagonal, only half the table needs to be shown. This fact is utilized by displaying the actual similarity values above the diagonal and the corresponding shade diagram below the diagonal: the shade diagram gives an immediate impression of the pattern of similarities, and these may be investigated further by examining the data values. Where possible, the use of colour can improve the visual impact of a display by using, say, increasing shades of red for values larger than some central value, and shades of blue for smaller values. For best results it is recommended that no more than four shading categories be used for monochrome printing, although this might be increased to about seven categories if different colours are used.

As an alternative to shading, the size of each cell value can be indicated by the size of a symbol. Figure 2.2 displays the individual yields of nine grassland species when grown in pairs as a competition diallel. The actual data values are given in Table 6.8 where a fuller analysis is carried out. In Fig. 2.2 the species have been ordered according to their overall mean yields (column means). The row means give the associate effects for each species, which is an inverse measure of their competitive ability. The main body of the figure indicates the sign and magnitude of deviations of individual yields from a simple additive model for species and associate effects. It is clear that the two *Trifolium* species

Associate species	*Species yield (log g/m²)*									*Mean yield of associate*
	Li	*Ms*	*Ae*	*Tp*	*Dg*	*Th*	*Fp*	*Lc*	*Pp*	
Lolium italicum	+	+	.	.	.	.	.	o	O	7.22
Medicago sativa	o	+	.	.	.	.	o	O	+	7.95
Arrhenatherum elatius	+	+	.	.	.	.	O	.	.	8.17
Trifolium pratense	.	.	.	+	o	.	.	O	O	8.15
Dactylis glomerata	.	.	+	.	.	.	+	.	.	8.41
Trifolium hybridum	O	o	.	.	.	.	.	.	.	8.70
Festuca pratensis	.	.	.	.	+	+	.	.	.	8.47
Lotus corniculatus	o	O	o	.	.	.	.	+	+	8.79
Phleum pratense	o	o	o	.	.	.	+	+	+	8.66
Mean yield of species	9.80	9.09	9.06	8.61	7.89	7.75	7.61	7.44	7.28	

Residual yields (log g/m²)

O	o	.	+	+
<−0.4	−0.4–−0.2	−0.2–0.2	0.2–0.4	>0.4

Fig. 2.2 Partition of yields of nine grassland species when grown in mixtures as a competition diallel. The symbols in the table indicate the deviations from the yields expected for a simple additive model for species and associate effects. The raw data values are given in Table 6.8.

fit this model well, whereas *Phleum* shows large residuals with most species, indicating very specific interactions. The contrasting pattern of column residuals for *Medicago* and *Lotus* is also of interest; *Medicago* appears to be relatively insensitive to competition from its associate species in mixtures whereas *Lotus* is particularly sensitive.

2.2 MAPPING

Data collected over a geographical region can be effectively displayed by superimposing the information on to a map. Data are usually only obtained at discrete locations and may be either mapped as discrete points, e.g. as for sightings of a rare species of bird, or shown as being continuous over a region, e.g. as the population density of a common species.

For the display of discretely located data, symbols of different size can be used to show a single variable. For example, in Fig. 2.3 the level of concentration of dieldrin residues in the livers of dead birds is shown by discs of different size centred on 10 km squares. (It appears from the figure that, if more than one dead bird were found in a square, the additional symbols are included in neighbouring squares.) In this example, extra information could be included on the map by using different symbols for the two species, e.g. circles for kestrels and triangles for owls, although the smaller symbols would be difficult to distinguish. Using symbols of different colours would be even more effective.

Mapping data that are continuous over a region can be done in several ways. The easiest is to divide the region into small regular areas, for example a square grid, and use some sort of shading scheme so that, for example, areas with higher population density are represented by darker shades. If the data have been sampled at regular intervals, the shaded area can then represent the value at a single location or the mean value of a set number of locations. For sample data which are irregularly spaced, values should first be interpolated to provide a data matrix based on a regularly spaced grid.

The production of maps is considerably speeded by computer. The adoption of a grid shading scheme is well suited to line-printer output, when the different densities of shading may be obtained by appropriate choice of symbols and overprinting. The final result is quite adequate for research use and, if care is taken over symbols and mapping intervals, may be satisfactory for publication (Coppock, 1976). Figure 2.4 maps the density throughout Great Britain of an aphid species for a week in June 1984, produced by the SYMAP V program (Laboratory of Computer Graphics, Harvard). A logarithmic scale was used in this example, with successively darker shading indicating a three-fold increase in density. Note that it is rather difficult to distinguish among the seven different shadings used: this may be compared with the clarity of Fig. 2.1.

An alternative form of display is the contour map. This is shown in Fig. 2.5

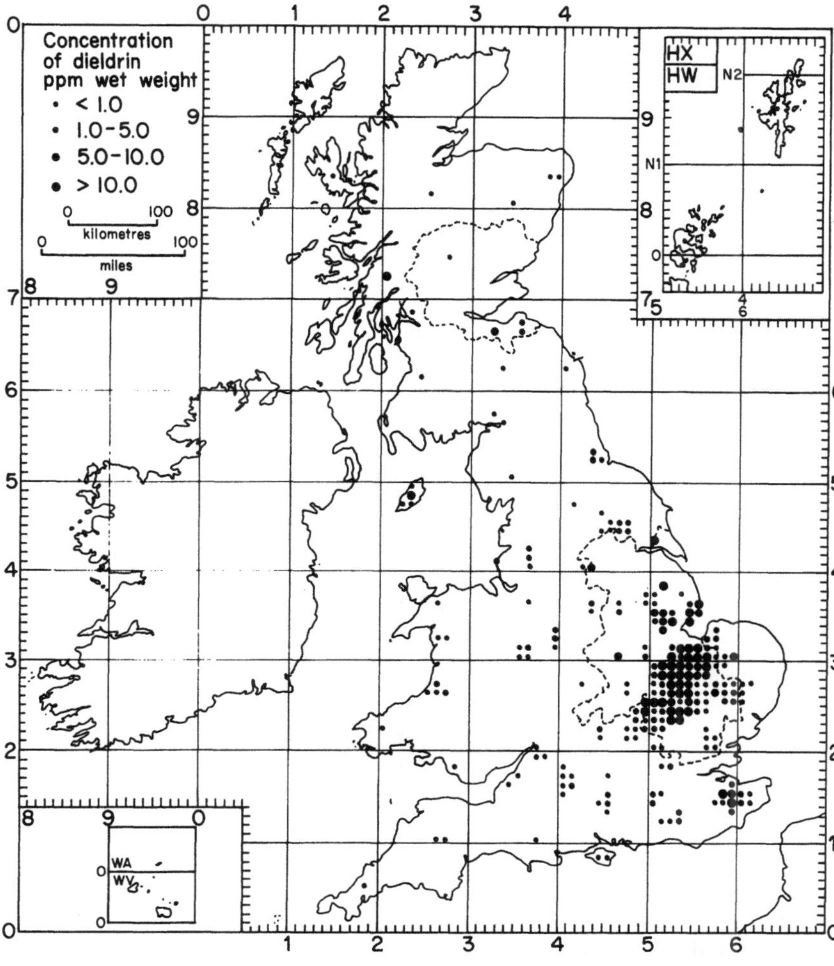

Fig. 2.3 Dieldrin residues in the livers of 227 kestrels and barn owls found dead during 1970–73. Each bird is represented by a point on the map. (Reproduced with permission from Fig. 9 of *Institute of Terrestrial Ecology Annual Report for 1974*.)

for the aerial density of the hop aphid, *Phorodon humuli*: the aphids' spatial distribution is seen to have two centres based on the primary hop growing regions of the UK. The map was produced by SURFACE II, a program developed by the Kansas Geological Survey (University of Kansas), and output to a high-quality graph plotter. When the density patterns are less distinctive, the usual impact of the contour maps may be improved by shading the area between contours. This can be done by a further processing of the results of SURFACE II followed by output to a dot-matrix printer, so

Fig. 2.4 Map of aerial density of *Sitobion avenea*, 11–17 June 1984 produced using the SYMAP program. Darker areas represent higher densities on a logarithmic scale (× 3 intervals). Numbers on map indicate positions of suction traps and their respective catch sizes ($\log_3$). (Reproduced with permission from Fig. 1 of Woiwod and Tatchell, 1984.)

producing a higher-quality shaded map: the results are particularly effective when colour printing is used (see, e.g., Taylor, 1986).

Contour maps can also be represented as three-dimensional perspective views (Fig. 2.6). These can be very effective in providing an immediate visual impression, although their success depends on an appropriate choice of position for the view.

2.3 DISPLAYING DISTRIBUTIONS OF VARIABLES

Traditionally, histograms have been used to display the distribution of variables. Tukey (1977) has shown how the information about the data values

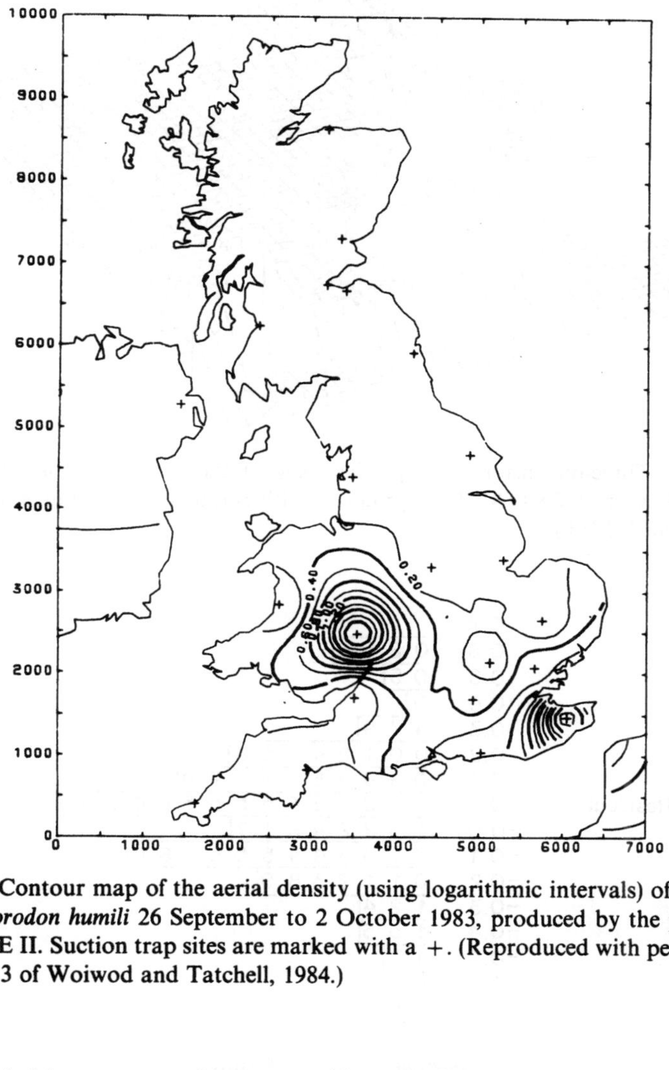

Fig. 2.5 Contour map of the aerial density (using logarithmic intervals) of the hop aphid *Phorodon humili* 26 September to 2 October 1983, produced by the program SURFACE II. Suction trap sites are marked with a +. (Reproduced with permission from Fig. 3 of Woiwod and Tatchell, 1984.)

in the basic histogram can be increased by using a stem-and-leaf plot. Figure 2.7 shows a stem-and-leaf plot used to display the deviations in yield from the sum of species and associated effects in Fig. 2.2. The class intervals for the histogram are printed vertically and form the 'stem', while the 'leaves' are the second digits of the individual values displayed horizontally within each class interval; the leaves are usually arranged in ascending order. Thus in Fig. 2.7(a) there are four residuals lying between 0.6 and 0.7, namely 0.60, 0.62, 0.65 and 0.66. Tukey (1977) describes several variants of the basic stem-and-leaf plot. One useful device is to place two plots back-to-back to compare the distribution of two variables: this is done for positive and negative residuals in Fig. 2.7(b). Using a stem-and-leaf display reduces the possibility of losing

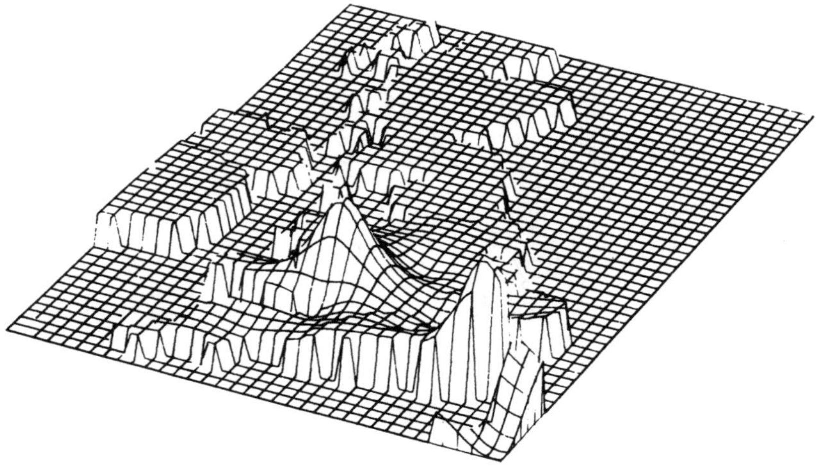

Fig. 2.6 Three-dimensional perspective view of the aphid densities in Fig. 2.5 obtained using SURFACE II. (Reproduced with permission from Fig. 4 of Woiwod and Tatchell, 1984.)

(a)

Residual	
0.7	
0.6	0 2 5 6
0.5	0 5
0.4	0
0.3	4 7 9
0.2	0 0 1 2 4 4 5 6
0.1	0 3 4 4 5 5 6 6 7 9
0.0	1 1 2 2 2 4 5 5 5 5 6 6 8 9 9
−0.0	1 1 2 2 4 6 7 7
−0.1	0 1 3 3 4 4 5 5 5 5 6 6
−0.2	0 1 1 2 3 8 9
−0.3	1 2 4
−0.4	1 5 5
−0.5	1 3
−0.6	6
−0.7	0 5

(b)

Negative		Positive
7 7 6 4 2 2 1 1	0.0	1 1 2 2 2 4 5 5 5 5 6 6 8 9 9
6 6 5 5 5 5 4 4 3 3 1 0	0.1	0 3 4 4 5 5 6 6 7 9
9 8 3 2 1 1 0	0.2	0 0 1 2 4 4 5 6
4 2 1	0.3	4 7 9
5 5 1	0.4	0
3 1	0.5	0 5
6	0.6	0 2 5 6
5 0	0.7	

Fig. 2.7 Stem-and-leaf plots of residuals from Fig. 2.2.
(a) overall distribution
(b) plots for positive and negative residuals placed back-to-back to examine the skewness of the overall distribution.

important detail of the distribution through an unfortunate choice of class boundaries. Thus in Fig. 2.8(a) we see that when the distribution of soil acidity across the Park Grass plots (Table 1.1) is displayed using a histogram with unit pH class boundaries, a possible division of plots into three groups is lost. However, with a stem-and-leaf plot this grouping may still be picked up from the distribution of second digits in Fig. 2.8(b); this is immediately apparent from Fig. 2.8(c).

Another idea for displaying distributions proposed by Tukey is the box-and-whisker plot. Again Tukey (1977) introduces several variants of differing complexities. Figure 2.9 displays the residual yields of Fig. 2.7 in this alternative way. The 'box' is drawn to cover the inter-quartile range (−0.15, 0.16) and is marked by a cross at the median of the distribution (0.02). The 'whiskers' usually extend from the box to the extremes of the distribution but, if some points are detected as obvious outliers, one variant of the plot is to

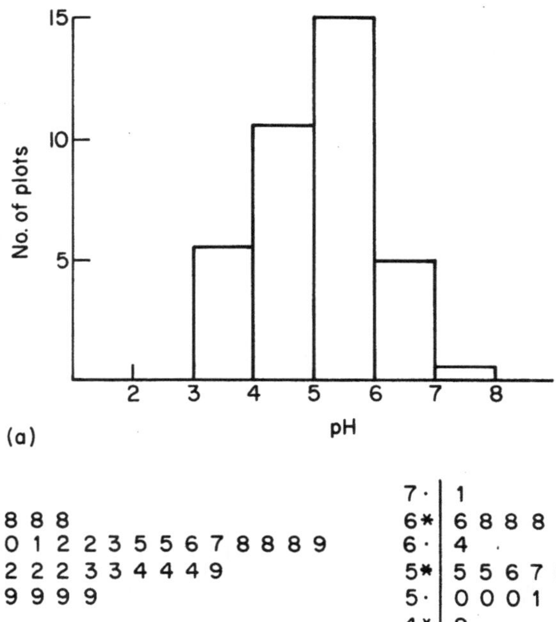

(a)

	7	1
	6	4 6 8 8 8
pH	5	0 0 0 1 2 2 3 5 5 6 7 8 8 8 9
	4	0 1 2 2 2 3 3 4 4 4 9
	3	7 8 9 9 9 9

(b)

7·	1
6*	6 8 8 8
6·	4
5*	5 5 6 7 8 8 8 9
5·	0 0 0 1 2 2 3
4*	9
4·	0 1 2 2 2 3 3 4 4 4
3*	7 8 9 9 9 9

(c)

Fig. 2.8 Distribution of soil pH values for the Park Grass plots (Table 1.1) using (a) standard histogram with unit class interval, (b) and (c) stem-and-leaf plots with class intervals of 1 and 0.5. In (c) each leading digit is represented by two adjacent points on the stem. For the lower point, e.g. 6·, only leaves 0 to 4 are shown, the upper point, e.g. 6*, includes leaves 5 to 9.

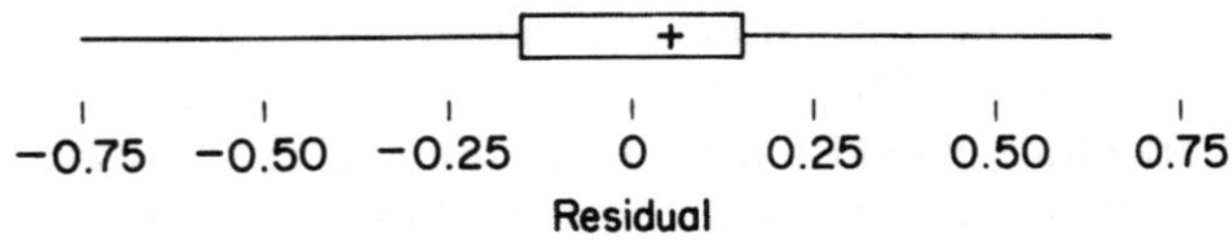

Fig. 2.9 Box-and-whisker plot of residuals in Fig. 2.7. The box covers the interquartile range and is divided at the median. The whiskers span the complete range of residuals.

extend the whiskers to the extremes of the 'reasonable' data values and mark the outliers by asterisks beyond the ends of the whiskers.

When several sets of data are being compared, the separate box-plots can be displayed side by side. Figure 2.10 shows the distribution of relative yields on the Park Grass plots displayed in this way for the eight grassland species which appear in more than 75% of the plots: a square-root scale has been used to give

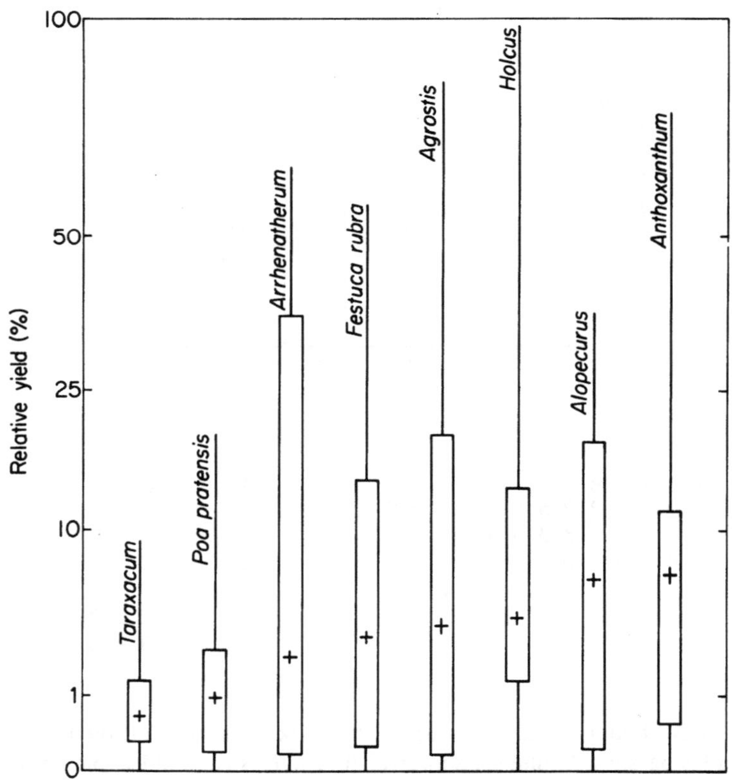

Fig. 2.10 Box-and-whisker plots for relative yields of eight species on the Park Grass plots arranged in increasing order of their medians.

a more symmetric distribution. The species have been ordered by the size of their median percentage yields: in contrast to ranking by mean yield, this ranking is independent of the choice of transformation. When the medians are all similar, arranging the individual box-plots by increasing length of box, i.e. inter-quartile range, may be revealing. In Fig. 2.10, the species show considerable variation in median percentage abundance and range. The species with lowest median abundance also had smallest inter-quartile range while *Arrhenatherum* had a low median abundance but the largest inter-quartile range extending from 0.05% to 37% of plot yield. When a data set is limited in size (say less than 30 units) it may be feasible to plot the individual data values along the box-plot, thus increasing the amount of information conveyed.

The previous example combines two types of distributional information, the frequency of occurrence of species on the Park Grass plots and their abundance when present. In many cases it is useful to separate these two essentially different types of information. Williams (1964) suggested plotting the total abundance of a species against the number of sites in which it is observed. Figure 2.11 shows his plot of the relation between abundance and frequency of occurrence for 15 beetles in 13 Canadian lakes. As an alternative to plotting the total abundance, the mean or median abundance for those sites in which the species appears could have been used; or the points for each species could be replaced by a box-and-whisker plot. For mobile species, their persistence at a site from year to year is also important. Figure 2.12 shows, for light-trap samples of moth species at 14 sites in the UK, the relationship between the number of sites in which a species is observed over a six-year period, and the average number of years for which it is observed at those sites. The average abundance of a species when it is observed in a year's sample at a site is indicated by the use of different symbols on the plot. The range of distribution of a species, its persistence and average abundance are all seen to be positively correlated but several species stand out from this trend. Species that occur infrequently at a large number of sites are most probably occasional migrants from the Continent; whereas species that occur persistently at only a proportion of sites are good indicator species for those sites. Species A, B and C are exceptionally good indicators: species A was found at just one site but occurred there in five years out of six; species B occurred in every year at three sites and nowhere else; species C occurred at the eight sites in southern England (see Fig. 3.12) in every year, except for one site in one year; all three species were fairly common or abundant when they appeared.

In several of the illustrations in this section we have been concerned with describing, for each species, the distribution of abundance across sites. An alternative characterization for species-by-sites data is given by considering, for each site, the distribution of abundance across species. Such distributions may be described by histograms, with class intervals chosen on a logarithmic scale, but other forms of display are useful, particularly when comparing the

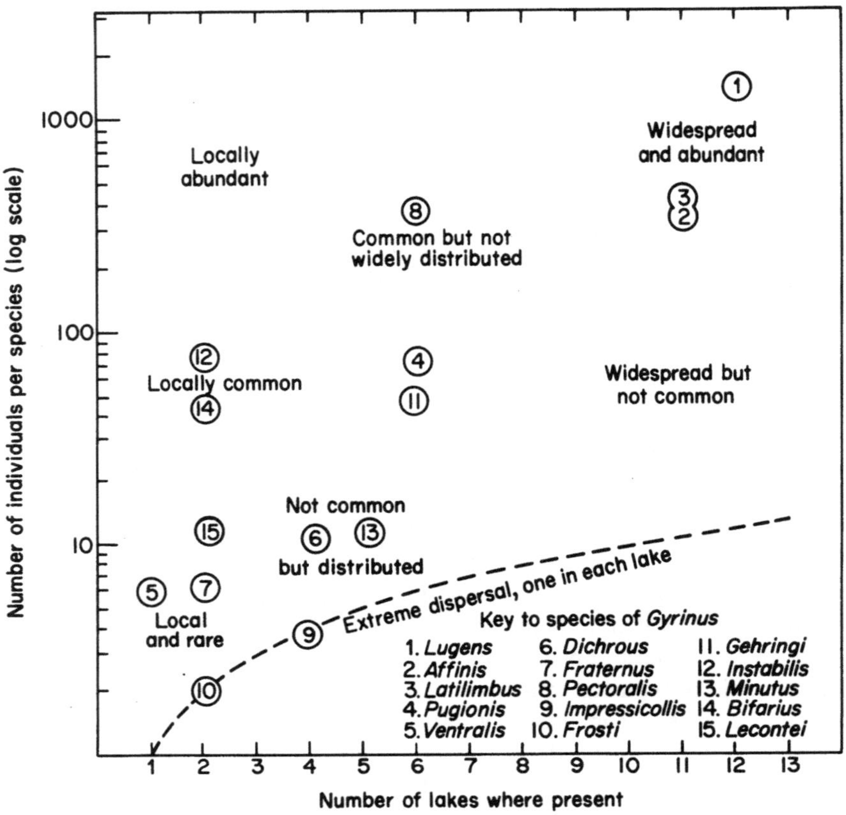

Fig. 2.11 Relation between abundance of individuals and frequency of occurrence of 15 species of Gyrinid beetles in 13 lakes in Mont Tremblant National Park, Montreal, Canada. (After Williams, 1964, Fig. 11.5.)

species diversities of sites. We briefly consider two graphical methods here: a third, the rarefaction curve, is illustrated in Section 3.8.1.

A commonly used way of displaying the pattern of relative abundances in a sample is the rank-size plot. Here the proportional abundances of the species are plotted against their ranks. Such a plot is used in Fig. 2.13 to show the progressive deterioration in floral richness that has occurred since 1856 on one of the Park Grass plots (plot 11/1d) after continuous application of a nitrogen fertilizer. As well as a dramatic reduction in the number of species, from 49 in 1856 to just 3 in 1949 (and 2 in 1973), there is a marked increase in the dominance of the commonest species, from 14.5% to 99.7%, and in the unevenness of the overall distribution of abundances.

A plot of cumulative abundances has advantages over the conventional rank-size plot since it provides the opportunity for identifying a diversity

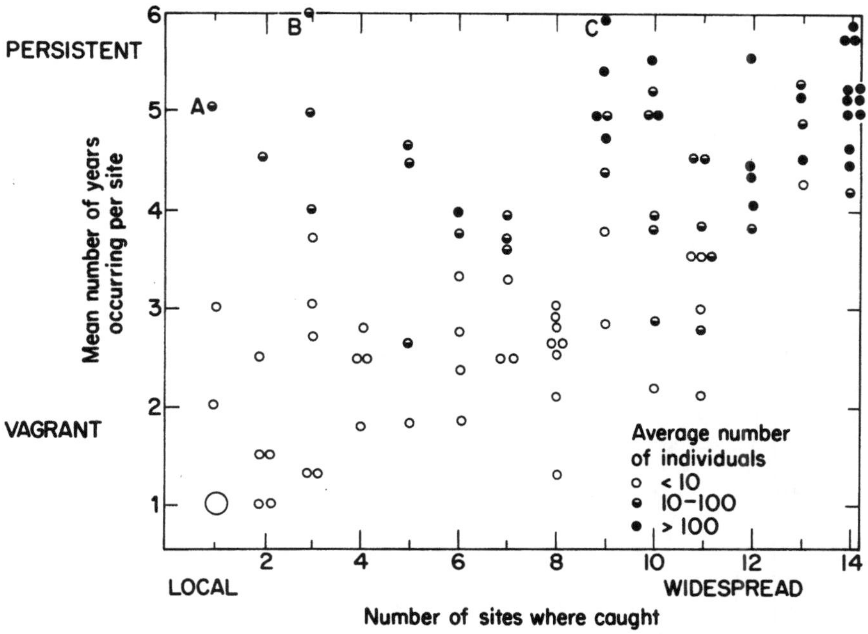

Fig. 2.12 Persistence and spatial distribution of moth species sampled at 14 sites over 6 years. A, B and C are identified as reliable indicator species for discriminating among sites.

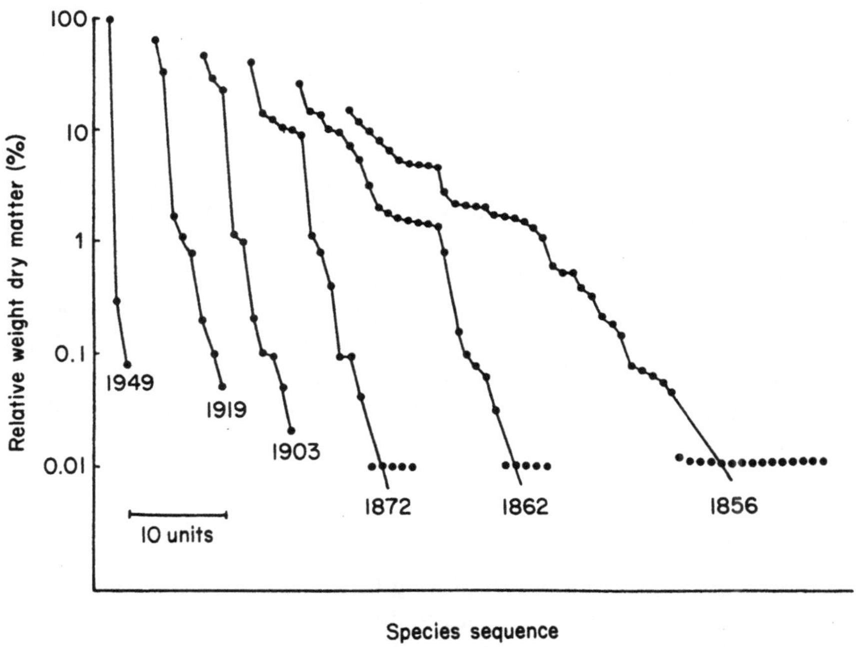

Fig. 2.13 Change in pattern of abundances of species on Park Grass plot 11/1d since 1856. Plots for separate years are translated horizontally to avoid overlap. Species recorded as having an abundance <0.01% are plotted as 0.01%. (Reproduced from Fig. 1 of Kempton, 1979.)

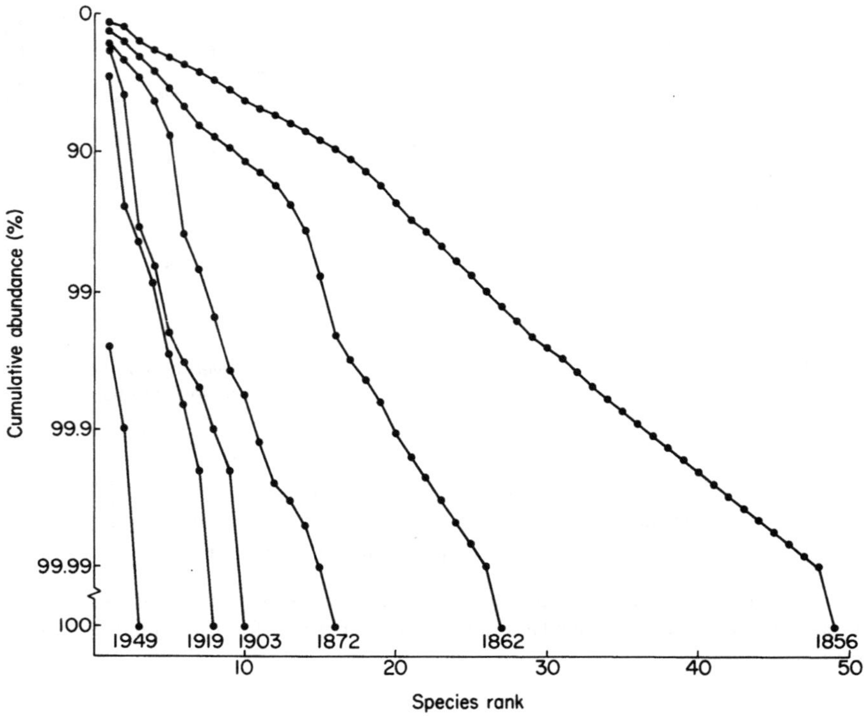

Fig. 2.14 Plot of the cumulative species frequency against rank for species abundance distributions on Park Grass plot 11/1d from 1856. Note the use of the complementary log scale.

ordering of the sites. Figure 2.14 plots, on a complementary log scale, the cumulative proportional abundance against rank for the Park Grass data of Fig. 2.13: using a log scale allows both rare and common species to be given similar emphasis on the plot. When, as in Fig. 2.14, the curves for separate samples do not intersect, an intrinsic diversity ordering of the samples exists which is followed by all common diversity measures (Solomon, 1979; Patil and Taillie, 1982). Thus for field plot 11/1d, there is an unequivocal and consistent reduction in species diversity since the initial application of nitrogen fertilizer in 1856.

2.4 BIVARIATE AND MULTIVARIATE DISPLAYS

The most common method for displaying bivariate data is the scatter plot. When the units to be displayed belong to different groups, the use of symbols of different shape, size or colour can be used to convey this information.

When there are more than two variables, one possibility is to produce all pairwise plots of the variables. Tukey and Tukey (1981) suggest that these plots be presented in a lower-triangular configuration as in Fig. 2.15. However, the number of displays to examine rises dishearteningly quickly as the number of variables increases, e.g. for 8 variables, 28 bivariate plots are required. Furthermore, any three- or higher-order relationships are obscured in the pairwise plots.

An alternative approach for many variables is the use of 'shapes'. In these displays, each individual is represented by one shape, and the different variables contribute to different parts of the shape. Several types of shape have been proposed; Fig. 2.16 (from Gower and Digby, 1981) shows a few examples

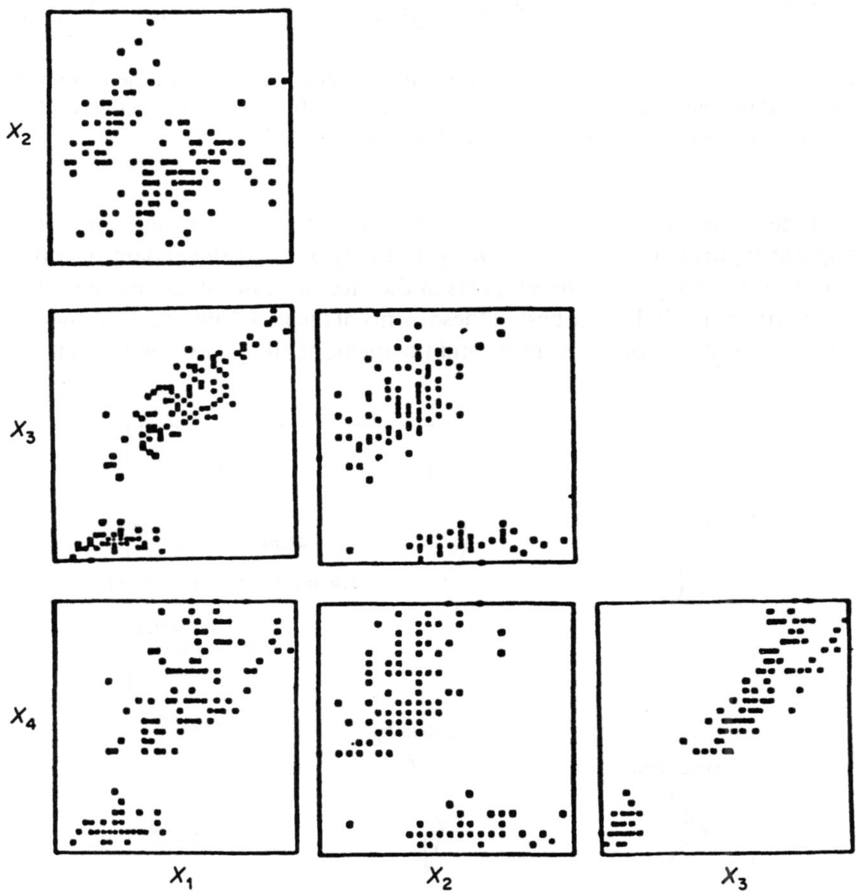

Fig. 2.15 Triangular arrangement of all pairwise scatter plots for four variables. Variables describe the length and width of sepals and petals for 150 iris plants, comprising 3 species of 50 plants. (Reproduced with permission from Fig. 10.12 of Tukey and Tukey, 1981.)

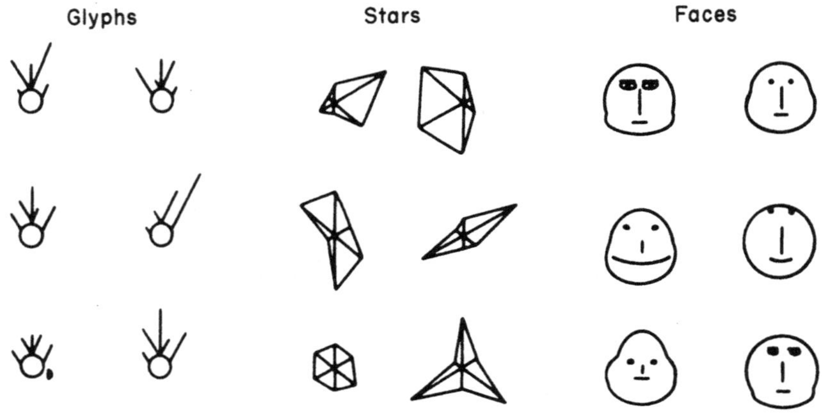

Fig. 2.16 Three types of shape for representing multivariate data. In these examples glyphs, stars and faces represent five, six and twelve (!) variables respectively. (After Gower and Digby, 1981. The faces are from Chernoff, 1973.)

of three forms, glyphs, stars and faces. The latter have achieved a certain popularity, probably for their novelty, but they are very dependent on which variables are assigned to which parts of the face: we cannot recommend their use. Stars are probably the best of these forms of display. Figure 2.17 shows the six stars for five Park Grass plots and the mean of all 38 plots, with each star

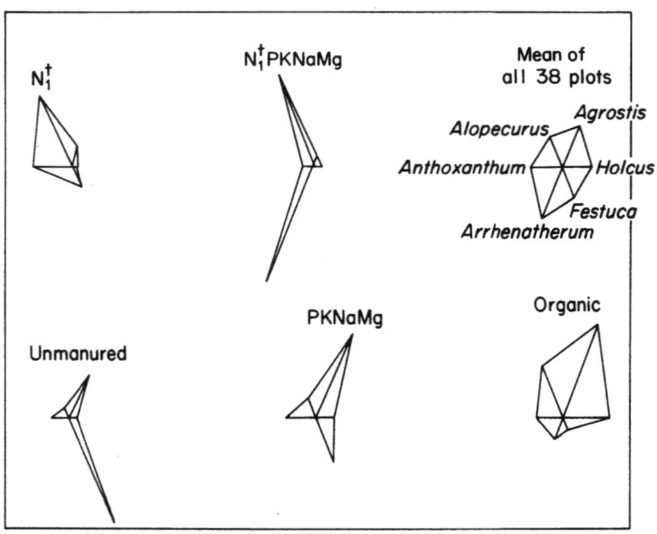

Fig. 2.17 Frequency of the six commonest species on the Park Grass plots using star displays.

comprising six arms which represent the relative abundances on the plot of the six commonest grass species. The star for the organic plot is seen to be most similar to the 'mean' star, while the stars for plot 3d (unmanured) and plot 16d ($N_1^{\dagger}$ PKNaMg) are very different, plot 3 having a high frequency of *Festuca* and *Agrostis* while the predominant species in plot 16 are *Alopecurus* and *Arrhenatherum*. The use of shapes can be effective for small sets of data, but when there are many shapes to compare it becomes difficult to spot the similarities and differences between them. However, they can be useful for conveying ancillary information when included with other types of display, as suggested previously.

The use of several different forms of display to convey information on a set of variables is demonstrated in Fig. 2.18. This illustrates how the occurrence of arctic-alpine vegetation groups is influenced by the amounts of calcium and phosphate available in the soil at sites on the cliffs of Snowdonia. Samples from soils with similar phosphate and calcium levels have been combined and plotted on the graph using circles of different sizes to indicate the abundance of vegetation. Finally, the circles have been segmented to form a pie chart which indicates the proportional abundance of each of seven vegetation classes. The combination of three different methods of display has here been used very effectively to illustrate the major patterns in the data.

The pie chart is very popular for displaying variables that sum to a constant value, typically one or 100%. However, for the case of three variables, these can often be displayed more clearly using a special type of scatter plot. The units can of course be displayed as points in three-dimensional space, but the constraint on their sum means that their variation is confined to a two-dimensional subspace; in fact the points must lie within an equilateral triangle with vertices whose three-dimensional coordinates are (1, 0, 0), (0, 1, 0) and (0, 0, 1). It is now quite easy to calculate the appropriate two-dimensional coordinates, known as barycentric coordinates, and plot the points within an equilateral triangle, as shown in Fig. 2.19.

As an example, consider Fig. 2.20 which displays the frequencies of genotypes for the MN blood group system in native populations from different regions of the world. The Aborigine populations on the far right of Fig. 2.20 are almost homozygous with the NN genotype approaching 100% of the population; in contrast, the points for the samples of Chinese are at the centre of the plot, indicating genotype frequencies roughly in the ratio MM:MN:NN=0.25:0.5:0.25, i.e. M and N genes are present in equal frequency. The genotype frequencies for all populations are seen to conform closely to those expected under Hardy–Weinberg equilibrium, i.e. if the frequency of M is p and the frequency of N is $q=1-p$, the equilibrium genotype frequencies are MM:MN:NN$=p^2:2pq:q^2$. Edwards (1972) uses barycentric coördinates to illustrate departures from Hardy–Weinberg equilibrium caused, for example, by differential viability among genotypes.

Finally, we consider three-dimensional perspective views as shown in Fig.

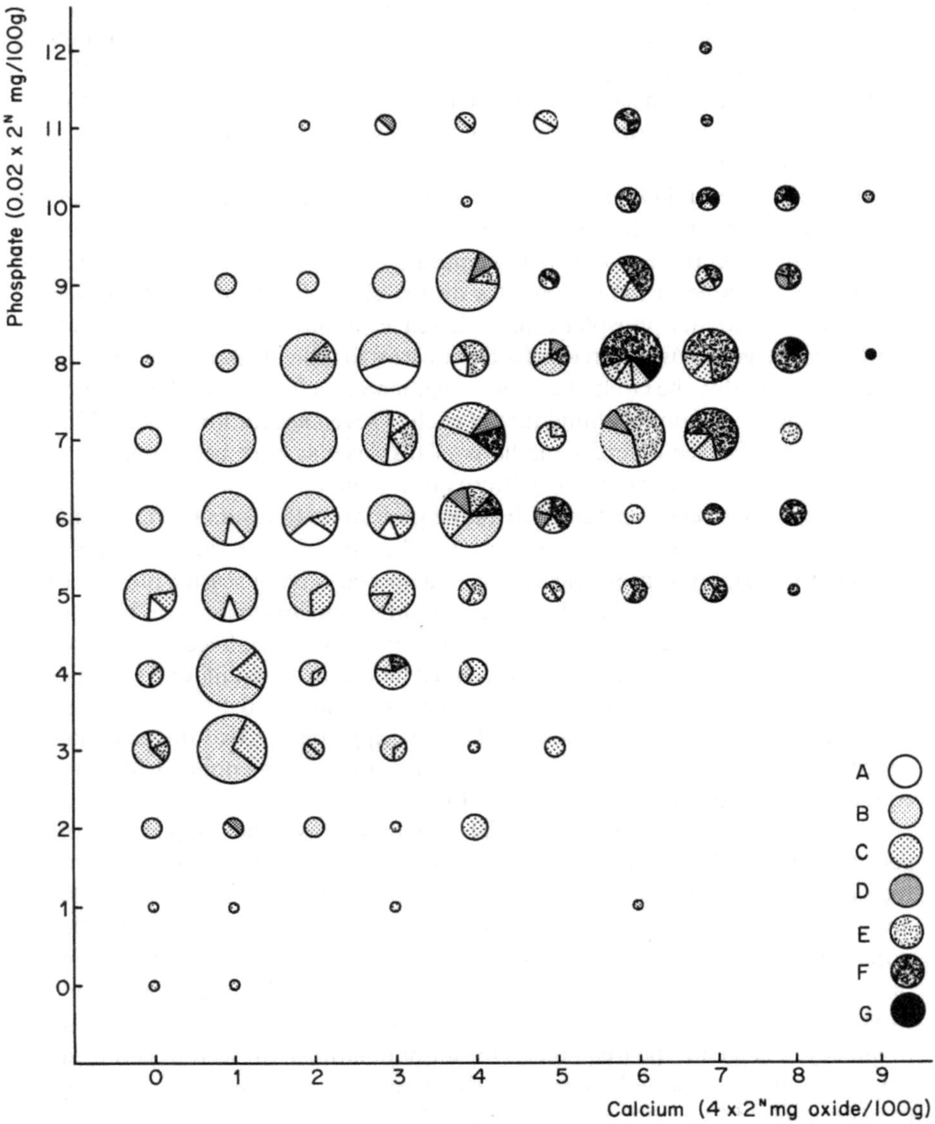

Fig. 2.18 Occurrence of seven vegetation groups at sites on cliffs of Snowdonia, from soils containing differing amounts of available phosphate and exchangeable calcium. The size of circles indicates the relative abundance of the vegetation. (Reproduced with permission from Plate 3 of *Institute of Terrestrial Ecology Annual Report for 1975*, where the vegetation groups are also described.)

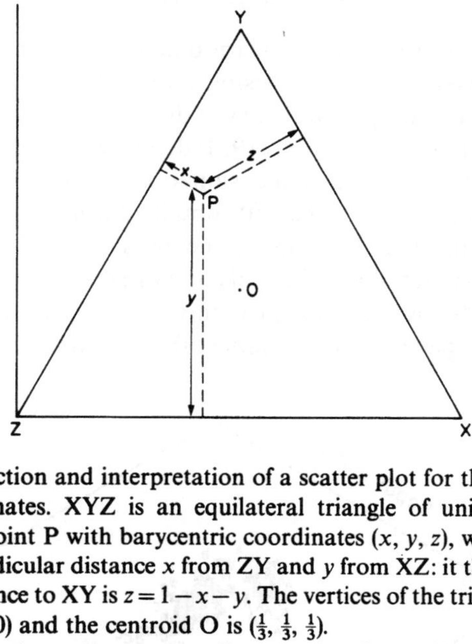

Fig. 2.19 Construction and interpretation of a scatter plot for three variables using barycentric coordinates. XYZ is an equilateral triangle of unit height (length of side $=2/\sqrt{3}$). The point P with barycentric coordinates (x, y, z), where $x+y+z=1$ is located at a perpendicular distance x from ZY and y from XZ: it then follows that the perpendicular distance to XY is $z=1-x-y$. The vertices of the triangle are X (1, 0, 0), Y (0, 1, 0), Z (1, 0, 0) and the centroid O is $(\frac{1}{3}, \frac{1}{3}, \frac{1}{3})$.

In two-dimensional coordinates the reference triangle can be constructed through the vertices $(2/\sqrt{3}, 0)$, $(1/\sqrt{3}, 1)$, $(0, 0)$; P is then plotted at $((2x+y)/\sqrt{3}, y)$.

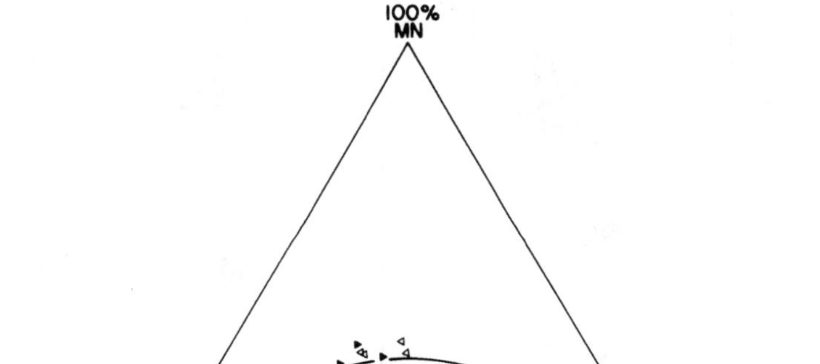

Fig. 2.20 Frequencies of genotypes of MN blood group system for native populations from different regions of the world. Solid squares, Aborigines (Australia and New Guinea); open triangles, Chinese; solid triangles, Indian; open circles, American Indians; solid circles, Eskimos (Greenland and Alaska). The curved line represent the expected pattern of frequencies for populations in Hardy–Weinberg equilibrium. (Data from Mourant, Kopec and Domariewska-Sobezak, 1976.)

2.21 for the first three variables of the iris data from Fig. 2.15. In Fig. 2.21 the variates are displayed on the three visible faces of the cube, as though it were viewed along a major diagonal: this view distorts all two-way plots equally. Green (1979) and Tukey and Tukey (1981) describe various other methods for direct views of three- and four-dimensional data, none of which are wholly successful. This emphasizes the benefits obtained when interpreting multivariate data, if the major variation among the units can be described by a single two-dimensional scatter plot. Principal components analysis, described in the next chapter, sets out to do just this by finding the plane in the multidimensional space which maximizes the variation among units.

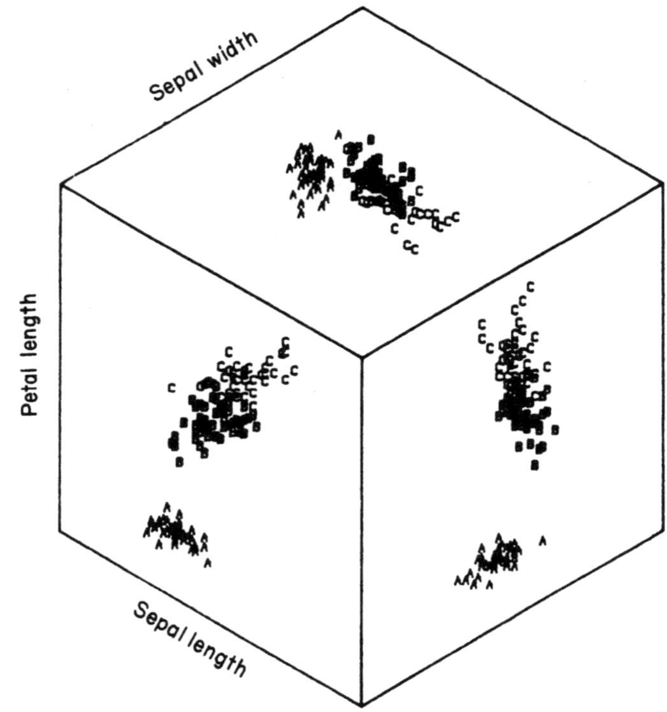

Fig. 2.21 Three-dimensional perspective view for the first three variables of the iris data (Fig. 2.15). Plants of the three species are coded A, B and C. (Reproduced with permission from Fig. 10.8 of Tukey and Tukey, 1981.)

3 Ordination

The term 'ordination' derives from early attempts to order a group of objects, for example in time or along an environmental gradient. Nowadays, the term is used more generally and refers to an 'ordering' in any number of dimensions (preferably few) that approximates some pattern of response of the set of objects. The usual objective of ordination is to help generate hypotheses about the relationship between the species composition at a site and the underlying environmental factors. With direct methods of ordination the experimenter must specify the environmental factors of interest and have independent knowledge of the score at each site for each factor (for a species ordination) or the species responses to each factor (for a site ordination). In contrast, most modern methods organize the data solely on the pattern of object responses and use any additional information on environmental variables only at a later stage to aid interpretation.

We shall first describe a direct method of ordination since it involves little computation and forms a useful introduction to the more objective methods developed later in this chapter, as well as giving an indication of the validity of these later methods. More extensive reviews of early ordination methods are given by Gauch (1982) and Greig-Smith (1983).

3.1 DIRECT GRADIENT ANALYSIS

We illustrate the method using the botanical analysis of the Park Grass plots introduced in Chapter 1. As the effect of the liming treatment has become of major interest in this experiment, it is natural to look first at the species response to the pH gradient of the plots.

In Fig. 3.1 the 38 plots have been grouped into six pH classes and the average frequency of the six dominant grass species plotted for each class. The species differ markedly in their response to soil reaction, the species *Arrhenatherum*, *Alopecurus* and *Festuca rubra* appearing more frequently in plots with neutral or high pH, while *Holcus*, *Anthoxanthum* and *Agrostis* appear relatively better adapted to more acid soils. Indeed, these latter three species dominate the flora on the most acidic plots (pH < 4), accounting for on average 95% of dry matter yield.

When working with a long environmental gradient, an ordination of species may be obtained using the position of each species' peak abundance along the gradient. Thus Fig. 3.1 might suggest a crude ordering *Anthoxanthum*, *Holcus*,

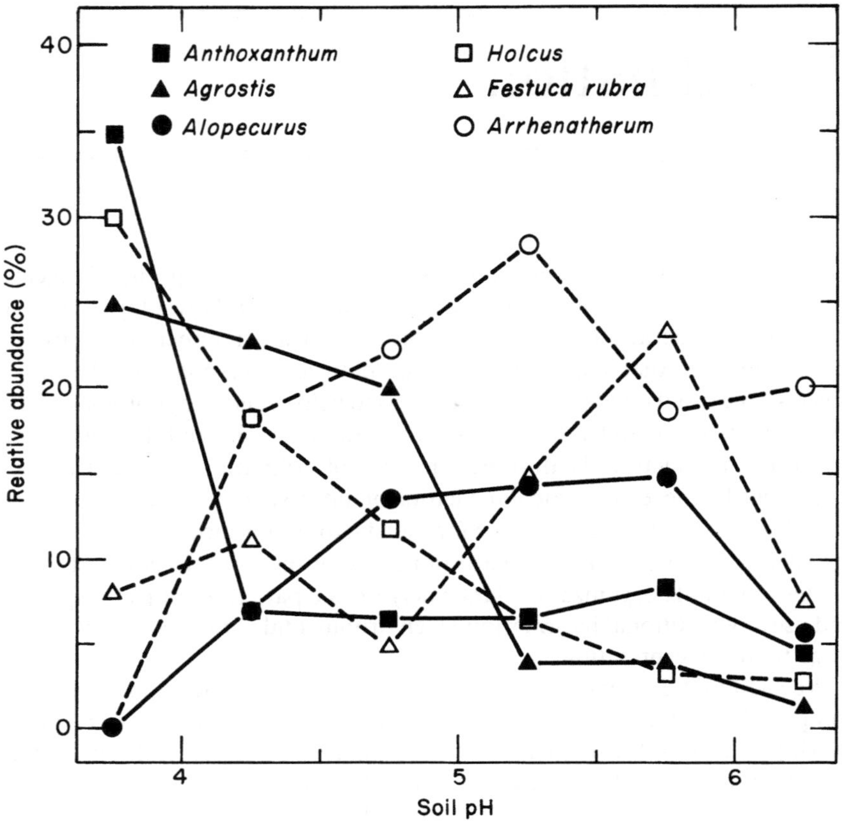

Fig. 3.1 Relative frequencies of six grass species on Park Grass plots in relation to soil pH.

Agrostis, *Arrhenatherum*, *Festuca*, with the position of *Alopecurus* indeterminate relative to the latter two species.

A more objective ordination procedure is to calculate a weighted score for each species based on its frequency in each environmental class. If the sites are classified into c classes and each class is given a weight w_j $(j=1, \ldots, c)$ to indicate its position along the environmental gradient, then species i with response x_{ij} in class j would have score

$$s_i = \sum_j x_{ij}w_j / \sum_j x_{ij}. \tag{3.1}$$

Using the six pH classes of Fig. 3.1 and choosing weights $w_j = j$ $(j=1, \ldots, 6)$, gives scores for the commoner Park Grass species shown in Table 3.1. For example, the score for *Agrostis* is 2.3, suggesting that this species may be used as an indicator of more acid soils; this agrees with Ellenberg's (1979)

Table 3.1 pH scores for commoner grassland species based on their distribution on Park Grass plots.

Species	*pH score*
Holcus lanatus	2.2
Agrostis tenuis	2.3
Anthoxanthum odoratum	2.4
Poa pratensis	3.3
Festuca rubra	3.8
Poa trivialis	3.9
Alopecurus pratensis	4.0
Arrhenatherum elatius	4.0
Dactylis glomerata	4.2
Lathyrus pratensis	4.4
Trifolium pratense	4.7
Plantago lanceolata	4.7
Taraxacum officinale	4.8
Heracleum sphondylium	4.9
Ranunculus acris et bulbosus	5.0
Anthriscus sylvestris	5.0
Leontodon hispidus	5.1
Poterium sanguisorba	5.3
Lolium perenne	5.6
Helictotrichon pubescens	5.7

observations of the species' distribution in Central Europe. However, this procedure will not distinguish between, say, species which are highly specific to neutral soils and species which are insensitive to soil reaction; both will have average scores.

Prior knowledge of species indicator values for a specific environmental factor allows one to reverse the above process and order sites using species indicator values as weights. Table 3.2 gives the average cover values for 50 species on 17 plots from a grazed meadow in Steneryd Nature Reserve, Sweden (Persson, 1981). For each species, an indicator value (Ellenberg, 1979) is given for the response to four different environmental factors: light, moisture, soil reaction and nitrogen. Scores may now be derived for each plot j and environmental factor k as a weighted sum of the species abundance scores x_{ij} $(i=1, \ldots, n)$ in that plot:

$$s_j(k) = \sum_i x_{ij} w_i(k) / \sum_i x_{ij}, \tag{3.2}$$

where the summation is over all species i for which the indicator value w_i for

Table 3.2 Abundance scores for species from plots in a park-meadow at Steneryd, south Sweden (Persson, 1981). Each score is the sum of cover values (0–5 scale) for nine sample quadrats. Species indicator values are from Ellenberg (1979).

Indicator values					*Plot number*																
L	M	R	N	*Species*	1	2	3	4	5	6	7	8	9	10	11	12	13	14	15	16	17
9	3	2	1	*Aira praecox*	9	1	.	.	.	.	.	.	.	.	.	.	.	.	.	.	.
8	3	1	1	*Teesdalia nudicaulis*	9	3	.	.	.	.	.	.	.	.	.	.	.	.	.	.	.
9	3	2	1	*Rumex tenuifolius*	9	2	6	.	.	.	.	.	.	.	.	.	.	.	.	.	.
5	–	–	–	*Vicia angustifolia*	4	6	8	1	.	.	.	.	.	.	.	.	.	.	.	.	.
6	5	5	4	*Trifolium dubium*	.	.	9	1	.	.	.	.	.	.	.	.	.	.	.	.	.
7	3	–	2	*Viscaria vulgaris*	1	17	.	5	.	.	.	.	.	.	.	.	.	.	.	.	.
7	4	7	3	*Galium verum*	5	6	12	4	3	.	.	.	.	.	.	.	.	.	.	.	.
7	4	–	2	*Hieracium pilosella*	12	7	16	8	1	6	.	.	.	.	.	.	.	.	.	.	.
8	4	–	5	*Achillea millefolium*	1	9	16	9	5	2	.	.	.	.	.	.	.	.	.	.	.
6	4	4	–	*Stellaria graminea*	1	4	9	.	4	3	.	.	.	.	.	.	.	.	.	.	.
7	3	3	–	*Festuca ovina*	38	43	43	30	10	11	20	.	.	5	4	1	1	.	.	.	.
6	–	–	–	*Plantago lanceolata*	2	9	7	15	13	8	.	.	.	.	.	.	.	.	.	.	.
8	–	–	7	*Trifolium repens*	.	.	6	14	19	2	.	.	.	.	.	.	.	.	.	.	.
7	–	–	–	*Trifolium pratense*	.	.	.	10	8	.	.	.	.	.	.	.	.	.	.	.	.
7	5	–	7	*Taraxacum* sp.	1	3	5	.	11	6	.	.	.	.	.	.	.	.	.	.	.
8	3	7	3	*Ranunculus bulbosus*	.	1	9	8	10	9	.	.	.	.	.	.	.	.	.	.	.
7	4	–	2	*Campanula rotundifolia*	.	5	8	1	.	7	.	.	.	.	2	.	.	.	.	.	.
6	5	–	5	*Cerastium fontanum*	.	.	2	7	9	4	.	.	.	.	.	.	.	.	.	.	.
7	4	8	2	*Filipendula vulgaris*	.	7	.	1	9	9	.	.	.	.	.	.	.	.	.	.	.
7	4	3	2	*Luzula campestris*	4	10	10	9	7	6	9	.	.	2	1	.	2	.	1	.	.
8	5	–	4	*Cynosurus cristatus*	.	.	.	9	18	9	.	.	.	.	.	.	.	.	.	.	.
6	6	–	6	*Alchemilla* sp.	.	.	.	.	14	7	.	.	.	.	.	.	.	.	.	.	.
7	–	3	3	*Agrostis tenuis*	10	12	19	15	16	9	.	9	28	8	.	4	.	.	.	.	.
–	–	5	–	*Anthoxanthum odoratum*	.	.	5	8	7	6	7	.	1	5	.	.	.	.	.	.	.
–	4	5	3	*Saxifraga granulata*	.	5	3	9	12	9	.	1	7	4	5	1	1	1	3	.	.

8	6	3	2	*Hypericum maculatum*		.	.	.	.	4	17	.	.	8	.	.	1	1	.	.	.	.
7	6	7	6	*Lathyrus pratensis*		.	.	.	4	.	5	.	.	.	9	.	.	.	.	.	.	.
8	–	5	5	*Rumex acetosa*		.	7	.	10	9	9	3	9	8	9	2	5	5	1	7	.	.
–	–	–	–	*Festuca rubra*		.	.	.	.	15	6	.	18	1	9	.	.	2	.	.	.	.
5	4	–	4	*Convallaria majalis*		.	.	.	.	.	.	10	.	.	.	.	.	.	.	.	.	.
6	5	–	–	*Poa pratensis*		1	.	5	6	2	8	10	15	12	15	4	5	6	7	.	.	.
6	4	–	–	*Veronica chamaedrys*		.	.	1	4	6	9	9	9	11	11	6	5	4	1	7	.	.
–	5	3	2	*Lathyrus montanus*		.	.	.	.	.	7	9	2	12	6	3	8	.	.	.	.	.
6	–	2	3	*Deschampsia flexuosa*		.	.	.	.	.	.	30	.	14	3	8	.	3	3	.	.	.
5	4	8	3	*Campanula persicifolia*		.	.	.	.	2	6	3	.	6	5	3	9	3	2	7	.	.
5	5	3	–	*Viola riviniana*		.	.	.	.	.	4	1	4	2	9	6	8	4	1	6	.	.
7	5	–	8	*Anthriscus silvestris*		.	.	.	.	1	6	.	.	.	.	7	.	.	7	3	1	.
5	5	6	5	*Stallaria holostea*		.	.	.	.	.	6	8	21	39	31	7	12	.	16	11	6	9
7	5	–	6	*Dactylis glomerata*		.	.	.	.	.	8	.	14	2	14	3	9	8	7	7	2	1
–	–	5	–	*Anemone nemorosa*		.	.	.	4	10	7	21	14	13	19	20	19	6	10	12	14	21
4	4	7	4	*Hepatica nobilis*		.	.	.	.	.	8	.	4	.	6	2	10	6	.	2	7	.
7	4	8	3	*Primula veris*		.	.	.	.	.	.	.	.	.	7	.	11	5	.	.	.	.
–	–	–	–	*Allium* sp.		.	.	.	.	2	7	.	1	.	3	1	6	8	2	.	7	4
5	5	5	3	*Poa nemoralis*		.	.	.	.	.	.	.	.	.	.	.	2	8	.	.	.	.
4	5	6	7	*Moehringia trinervia*		.	.	.	.	.	.	.	.	.	.	.	2	23	.	.	.	.
4	–	7	7	*Fraxinus excelsior (juv.)*		.	.	.	.	.	8	.	.	6	5	4	7	9	8	8	7	6
4	5	–	7	*Geum urbanum*		.	.	.	.	.	7	.	2	2	1	.	7	9	2	3	8	7
6	5	7	7	*Veronica hederifolia*		.	.	.	.	.	.	.	.	.	.	.	.	.	9	3	1	.
4	6	7	7	*Ranunculus ficaria*		.	.	.	.	.	.	.	.	.	.	13	.	.	21	20	21	37
2	–	7	7	*Mercurialis perennis*		.	.	.	.	.	.	.	.	.	.	1	.	.	.	11	45	45
				Plot score	Light	7.3	7.0	7.0	7.1	7.1	6.5	6.1	6.0	6.0	6.0	5.5	5.5	5.2	5.2	4.9	3.3	3.3
					Moisture	3.3	3.5	3.8	3.9	4.4	4.6	4.1	4.8	4.8	4.7	4.8	4.6	4.7	5.2	5.0	5.3	5.7
					Reaction	2.8	3.7	4.0	4.2	4.8	5.0	3.6	5.0	4.5	5.3	5.1	5.7	5.7	6.2	6.3	6.7	6.6
					Nitrogen	1.9	2.9	3.4	3.9	4.4	4.1	3.2	4.9	3.8	4.6	5.2	4.5	5.5	6.2	6.0	6.9	6.8

that environmental factor is known. For example, the nitrogen score for the first plot is formed as

$$\frac{9\times1+9\times1+9\times1+1\times2+5\times3+12\times2+1\times5+1\times7+4\times2+10\times3}{9+9+9+1+5+12+1+1+4+10}=1.9.$$

In this case the ordination of plots (Table 3.2) is very similar for all environmental factors, the main difference between plots being in the amount by which they are shaded by trees (Persson, 1981).

The reciprocal relationship between the species scores derived from environmental weightings by equation (3.1), and environment scores derived from species weightings (3.2), suggests a method for obtaining a refined ordination of sites. Species scores are first obtained using crude environmental values, perhaps for only a proportion of the sites, and then these species scores are used as weights to derive a refined site ordination. If this method is repeated, the site and species scores are each found to converge to a unique solution which is independent of the starting values. This procedure has been termed 'reciprocal averaging' by Hill (1973) and provides an objective ordination technique which will be discussed fully in Section 3.3.1.

While direct gradient methods have been used imaginatively to define and refine both species and site ordinations (see Gauch, 1982 for an extensive review), they rely on ancillary information about environmental gradients and require some subjective judgement on the part of the experimenter. Even when a dominant environmental gradient has been properly identified, failure to identify secondary gradients may lead to false interpretation. Thus Fig. 3.1 suggests that the species *Holcus*, *Anthoxanthum* and *Agrostis* might be found in equal abundance on the most acidic Park Grass plots; but examination of the seven unlimed plots with the lowest pH (Table 3.3) indicates that there is only limited coexistence. A second environmental gradient of plot productivity appears to be important in determining which species will be dominant, with *Holcus* doing best on the high-yielding plots while *Agrostis* is relatively better adapted to the poorer soils.

Table 3.3 Abundance (% total dry matter) of three dominant grass species on the most acidic unlimed sub-plots at Park Grass in 1973.

	Plot number						
Species	11^1	11^2	9	4^2	10	18	1
Agrostis	0	2	15	24	31	83	84
Anthoxanthum	5	0	72	76	69	17	11
Holcus	95	98	13	0	0	0	0
pH	3.7	3.8	3.9	3.9	3.9	3.9	4.1
Yield (t ha^{-1})	4.7	7.0	4.5	2.6	3.1	1.1	0.6

The more objective ordination methods which we shall now be considering rely only on the table or matrix of responses to define an ordering. Interpreting the orderings of the species and sites in successive dimensions in terms of environmental gradients may be carried out, after the main analysis. The methods of analysis are developed in terms of matrix algebra: readers unfamiliar with the terminology should refer to the Appendix.

3.2 PRINCIPAL COMPONENTS ANALYSIS

Principal components analysis (PCP) is the oldest of the indirect ordination methods. (We use the initials PCP here, rather than the commonly used acronym PCA, to avoid confusion with the more recent technique of principal coordinates analysis (PCO) which we describe in Section 3.5.) Suppose data are available for n units (species) and p variables (sites). Then we represent the response for the ith unit and jth variable as x_{ij} and the full set of data as the $(n \times p)$ matrix $\mathbf{X}$ with n rows and p columns. The purpose of the analysis is to produce an ordination of the units in a small number of dimensions which emphasizes the major patterns of variation in their responses.

In common with many ordination methods, the data values x_{ij} are initially standardized so that each variable has zero mean and, optionally, unit variance. Thus, if $x_{.j}$ is the mean and s_j^2 the variance of the jth variable, we transform x_{ij} to

$$y_{ij} = (x_{ij} - x_{.j}) \tag{3.3}$$

to set each variable to have zero mean, or as

$$y_{ij} = (x_{ij} - x_{.j})/s_j \tag{3.4}$$

so that each variable also has unit variance.

There are now two different, but equivalent, ways of considering PCP which give rise to two different, but equivalent, computational approaches. We consider the matrix $\mathbf{Y}$ $(n \times p)$, with transformed values y_{ij}, as giving the coordinates of n points in p dimensions, e.g. the first unit is at $(y_{11}, y_{12}, \ldots, y_{1p})$. For any specified $k < p$, PCP finds the subspace of k dimensions for which the sum of squares of perpendicular distances from the n points to that subspace is a minimum. Each point is then represented by the projection of its original position onto that subspace. For example, Fig. 3.2 gives a graphical representation of 10 points in two dimensions for the simple transformed matrix $\mathbf{Y}$ in Table 3.4. The best-fitting one-dimensional subspace is the line L and the PCP solution is the projection of the points onto the line L, giving the new set of points indicated by circles. The alternative representation of PCP seeks the k-dimensional subspace which maximizes the variation of the units within that space. In this case the line L is chosen to maximize the variation among the circled points. We shall now outline the computations arising from these alternative approaches before showing that they give equivalent results.

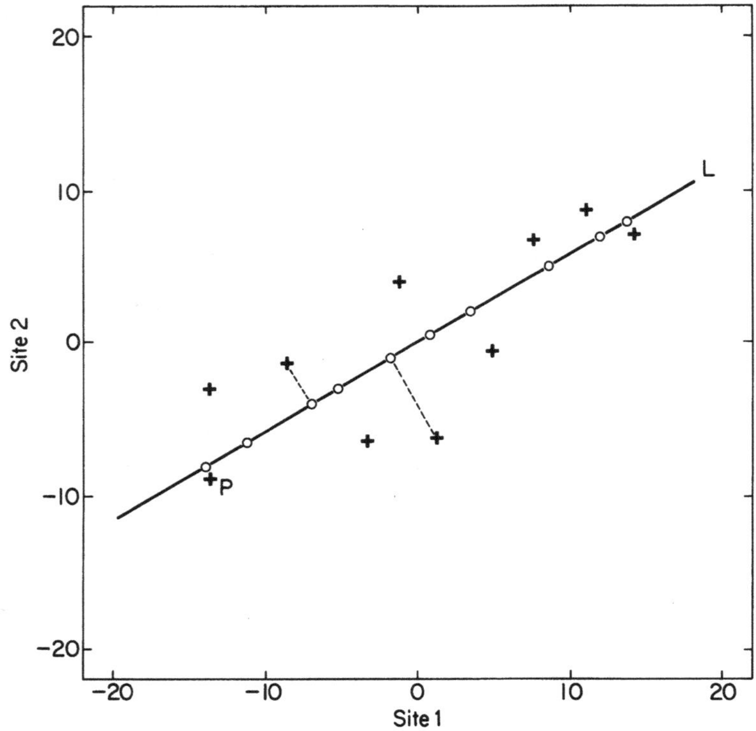

Fig. 3.2 Graphical representation of the species-by-sites matrix in Table 3.4. + indicates the coordinates of the ten species, open circles their projections onto the principal axis L.

3.2.1 Theory and computations

To follow the first approach, we consider the singular value decomposition (SVD) of the transformed data matrix **Y**:

$$\mathbf{Y} = \mathbf{USV}', \tag{3.5}$$

where **U** is an $(n \times p)$ orthonormal matrix, **S** is a diagonal matrix of order p, and **V**′ is the transpose of a $(p \times p)$ orthogonal matrix **V**. Since **U** is orthonormal, $\mathbf{U}'\mathbf{U} = \mathbf{I}$, the identity matrix of order p; likewise, because **V** is orthogonal $\mathbf{VV}' = \mathbf{V}'\mathbf{V} = \mathbf{I}$. (For further details see the Appendix.) It is assumed here that $n \geqslant p$; when this is not the case similar results hold for the SVD of **Y**′. The diagonal elements of **S** are called the singular values of the decomposed matrix **Y**, and are always arranged so that $s_1 \geqslant s_2 \geqslant, \ldots, \geqslant s_p \geqslant 0$.

The matrix **V** defines a rotation of the original axes to a new set of axes (the principal axes). The rotation is applied to the data by postmultiplying the matrix **Y** by **V** to obtain the coordinates of the points for the units relative to

Table 3.4 Singular value decomposition of a 10 × 2 species-by-sites matrix **Y**.

$$\underset{\mathbf{Y}}{\begin{bmatrix} -13.36 & -8.87 \\ -13.26 & -3.04 \\ -8.43 & -1.40 \\ -3.20 & -6.46 \\ 1.27 & -6.20 \\ -1.13 & 3.96 \\ 4.96 & -0.60 \\ 7.66 & 6.73 \\ 11.12 & 8.73 \\ 14.36 & 7.13 \end{bmatrix}} = \underset{\mathbf{U}}{\begin{bmatrix} -0.483 & -0.094 \\ -0.392 & 0.378 \\ -0.241 & 0.283 \\ -0.181 & -0.378 \\ -0.060 & -0.567 \\ 0.030 & 0.378 \\ 0.121 & -0.283 \\ 0.302 & 0.189 \\ 0.423 & 0.189 \\ 0.483 & -0.094 \end{bmatrix}} \underset{\mathbf{S}}{\begin{bmatrix} 33.14 & 0 \\ 0 & 10.58 \end{bmatrix}} \underset{\mathbf{V}'}{\begin{bmatrix} 0.867 & 0.500 \\ -0.500 & 0.867 \end{bmatrix}'}$$

Y represents the coordinates of the 10 sites on the original axes of Fig. 3.2, while the coordinates on the principal axes **A** = **US** are:

$$\begin{bmatrix} -16 & -1 \\ -13 & 4 \\ -8 & 3 \\ -6 & -4 \\ -2 & -6 \\ 1 & 4 \\ 4 & -3 \\ 10 & 2 \\ 14 & 2 \\ 16 & -1 \end{bmatrix}$$

their principal axes; these coordinates are often termed 'scores'. Thus, if a_{ij} is the score for the ith unit along the jth principal axis, it is given by

$$a_{ij} = y_{i1}v_{1j} + y_{i2}v_{2j} + \ldots + y_{ip}v_{pj} = \sum_{l=1}^{p} y_{il}v_{lj} \tag{3.6}$$

and the $(n \times p)$ matrix of scores, **A** is **YV**. Now consider the SVD of **Y**: we have **A** = **YV** = **USV′V**, whence, since **V** is orthogonal (**V′V** = **I**), we obtain **A** = **US**.

So far there has been no reduction in dimensionality. The original points were in p dimensions and the scores are also given in p dimensions as the p columns of the $n \times p$ matrix **A**; only the axes have changed. To obtain the best-fitting subspace of k dimensions, we take the first k columns of scores from **A**. Subscripting the matrices to indicate their number of rows and columns (for **S**) or columns (for **A** and **U**), the scores for the best subspace are given by $\mathbf{A}_k = \mathbf{U}_k\mathbf{S}_k$. Hence the PCP solution may be written down immediately once we have derived the singular value decomposition of the matrix **Y**.

As an example, consider the singular value decomposition for the 10 × 2 matrix given in Table 3.4. The first row of **A** gives the coordinates of the point P of Fig. 3.2, but referred to the line L as axis 1 and the line perpendicular to L through the origin as axis 2; these are termed the first and second principal axes. The point P is located 16 units along L and 1 unit at right angles to it. The matrix **V** is

$$\begin{bmatrix} \cos 30° & -\sin 30° \\ \sin 30° & \cos 30° \end{bmatrix}$$

and defines the rotation of the original axes through 30° to obtain the new axes.

We now consider an alternative approach to PCP and demonstrate the equivalence of the two methods. This more common approach entails finding a subspace of the transformed data matrix **Y** that maximizes the variation within the subspace. For this we need to introduce the sum of squares and products (SSP) matrix for **Y** given by $\mathbf{\Sigma} = \mathbf{Y}'\mathbf{Y}$.

We note first that, with some sets of data, the variables are measured on very different scales and it may then be preferable to analyse the correlation structure of the data rather than the SSP structure. This is handled quite neatly by the use of the second transformation of **X** to **Y** (equation (3.4)). In that case the columns of **Y** have unit variance and the SSP matrix $\mathbf{\Sigma}$ is simply the correlation matrix. Hence we shall proceed with the analysis of $\mathbf{\Sigma}$ and merely remember to use the second transformation if we want to analyse the correlation structure.

The analysis proceeds by finding the eigenvalues and eigenvectors of the $(p \times p)$ matrix $\mathbf{\Sigma}$. These are obtained from the spectral decomposition of $\mathbf{\Sigma}$:

$$\mathbf{\Sigma} = \mathbf{\Gamma}\mathbf{\Lambda}\mathbf{\Gamma}',$$

where $\mathbf{\Gamma}$ is a $(p \times p)$ orthogonal matrix, and $\mathbf{\Lambda}$ is a diagonal matrix of order p with diagonal elements $\lambda_1 \geqslant \lambda_2 \geqslant \ldots \geqslant \lambda_p \geqslant 0$. The λ_i $(i = 1, \ldots, p)$ are the eigenvalues of $\mathbf{\Sigma}$, arranged by convention in decreasing order of magnitude, while the columns of $\mathbf{\Gamma}$ are the corresponding eigenvectors. The coordinates of the points on the principal axes are now obtained by postmultiplying **Y** by the matrix $\mathbf{\Gamma}$, and the k-dimensional subspace that maximizes the variation within itself is then given by the first k columns of $\mathbf{A} = \mathbf{Y}\mathbf{\Gamma}$.

These two approaches to PCP, via the singular value decomposition of **Y** or the spectral decomposition of $\mathbf{Y}'\mathbf{Y}$, are closely linked mathematically. Indeed, we show in the Appendix that the eigenvectors of $\mathbf{Y}'\mathbf{Y}$ are simply the columns of the orthogonal matrix **V** in equation (3.5), i.e. $\mathbf{\Gamma} = \mathbf{V}$, while the eigenvalues of $\mathbf{Y}'\mathbf{Y}$ are the squares of the singular values of **Y**, i.e. $\mathbf{\Lambda} = \mathbf{S}^2$.

The eigenvalues, $\lambda_1, \lambda_2, \ldots, \lambda_p$, may be used to indicate the goodness-of-fit of the subspace of k dimensions. It can be shown that the variation in the lth dimension is given by λ_l; writing

$$t_k = \sum_{l=1}^{k} \lambda_l,$$

the total variation in the data values y_{ij} is t_p; the 'percentage variance explained' by the k-dimensional subspace is $100t_k/t_p$; the residual or unexplained variation is $(t_p - t_k)$, which itself can be expressed as a percentage of t_p. Of course, knowing that $\lambda_l = s_l^2$, these statistics can also be calculated from the singular value decomposition of **Y** that we adopted earlier. For example, in Table 3.4 the percentage variance about the origin of the 10 points (species) in two dimensions (sites), which is described by their positions along the first axis (line L in Fig. 3.2), is given by $(33.14)^2/[(33.14)^2+(10.58)^2]=91\%$.

Having chosen a value of k, possibly after consideration of the relative sizes of the eigenvalues λ, and obtained the matrix of scores $\mathbf{A}_k$, the units may be displayed for visual inspection and interpretation of their pattern. If k is fairly small, ideally 2, but possibly 3 or 4, the dimensions can be plotted against each other in pairs; e.g. if $k=3$, separate plots of dimensions 1 and 2, 1 and 3, and 2 and 3 can be produced. Alternatively, extra dimensions can be indicated on a graph of the first two dimensions as described in Section 2.4.

3.2.2 Example of principal components analysis

We now use the PCP method to derive an ordination of the species on the Park Grass plots. The data matrix **X** (Table 1.1) now represents the relative abundance (% total plot yield) of 44 species (rows) on 38 plots (columns). Since the column totals are 100%, the transformation to zero mean (equation (3.3)) is simply carried out by subtracting 100/44 from all x_{ij}; the matrix of species scores **A** on the principal axes is then obtained as outlined above. The first five eigenvalues were calculated to be 29.6%, 26.5%, 18.2%, 11.3% and 8.7% of the total sum t_p; hence these five principal axes account for 94.4% of the total variation in the species relative abundances, an excellent representation of the original data matrix.

The individual principal axes provide separate ordinations of the species. Figure 3.3 shows the close relationship between species scores on axis 1 and their mean abundance per plot: this axis represents a size component and conveys no information on species distribution. Figure 3.4 shows the species positions on the next four principal axes, which together account for 92% of the remaining variation once the size component has been removed. The information from these two plots is seen to be almost wholly restricted to the six dominant grass species shown in Fig. 3.1. Different axes separate different groups of these species; for example, *Agrostis*, *Anthoxanthum* and *Festuca*, which appear close together on axes 2 and 3, are separated in the fourth and fifth dimensions. Hence the large proportion (92%) of the total variation in adjusted species relative abundances accounted for by these four dimensions is not so remarkable since the majority of the variation arises from only 6 of the 44 species, and the variation of these 6 species could be represented completely in just 5 dimensions (see Section 1.4.5).

The emphasis of PCP on the most abundant species may be reduced by initially transforming the relative abundances x to $\log(x+1)$, while the species size component may be removed, before applying PCP, by adjusting all

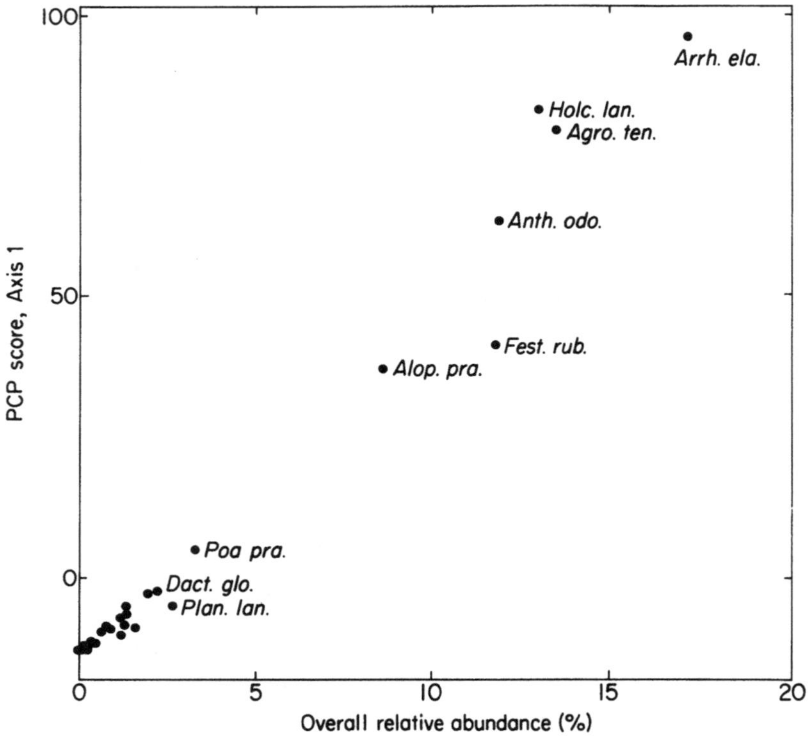

Fig. 3.3 First principal components scores for 46 species on the Park Grass plots (based on untransformed relative abundances) related to the overall relative abundance of the species.

species to have the same mean. Hence adjustment is now by both rows and columns, and equation (3.3) becomes

$$y_{ij} = x_{ij} - x_{i.} - x_{.j} + x_{..}, \tag{3.7}$$

where x_{ij} is now the log relative abundance of species i in plot j. The species scores on the first principal axis (Fig. 3.5) now show the six dominant grass species occupying similar positions to those on the second principal axis previously (Fig. 3.4), but much greater separation of species with moderate abundance. The second axis separates *Festuca* and *Holcus* among these dominant species and hence appears to combine information from axis 3 and 4 of the untransformed analysis. The first four eigenvalues are now 40.4%, 12.6%, 11.1% and 9.6%, of the total sum of squares; their sum is lower than before, emphasizing that, with more species now contributing, the pattern of variation is more complex. Interpretation of this two-dimensional ordination is rather limited without extra information. If two species (e.g. *Arrhenatherum* and *Alopecurus*) appear close together in Fig. 3.5, one might deduce that they

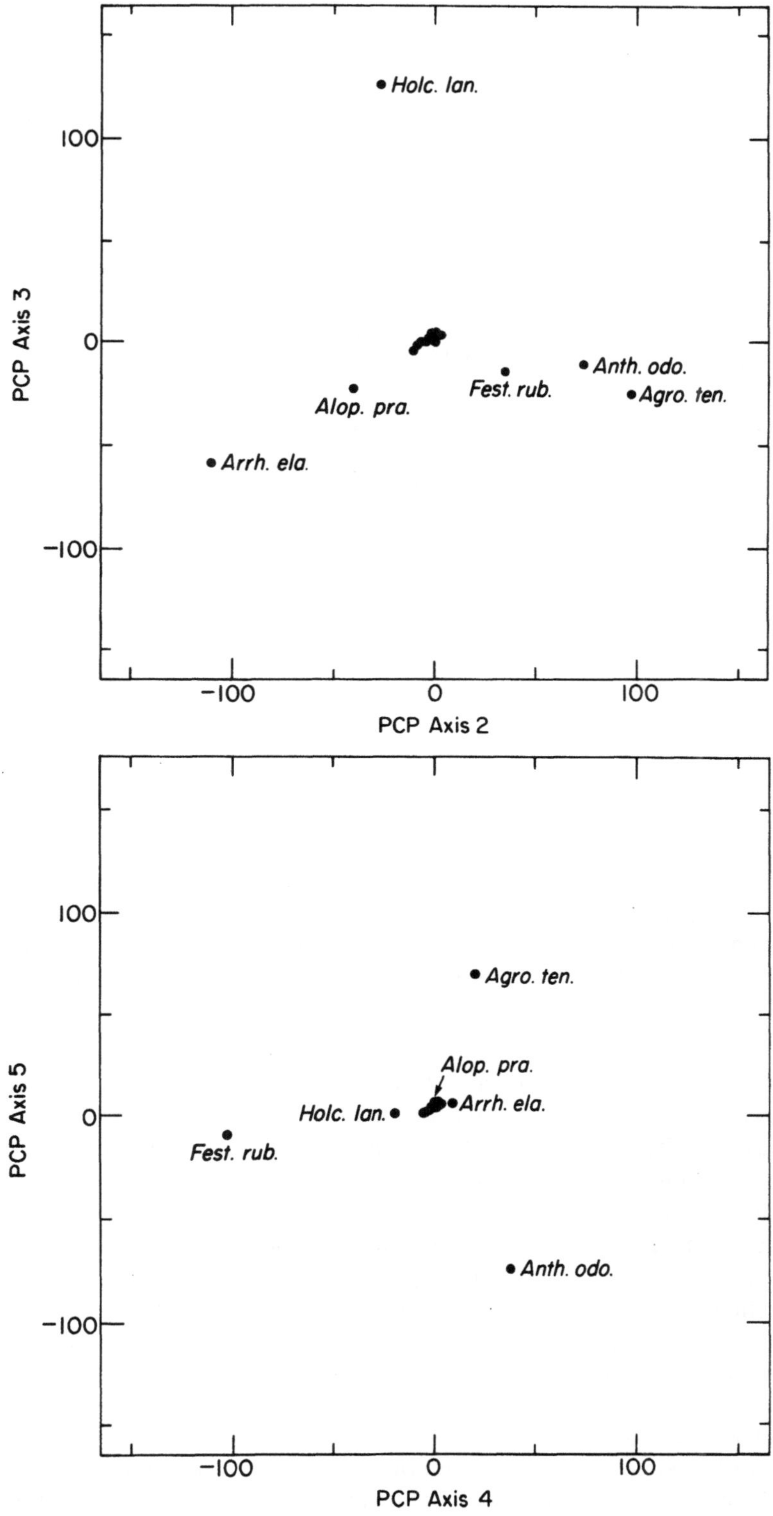

Fig. 3.4 Plots of scores for the Park Grass species on the principal components axes 2–5 (untransformed data). Species other than the six dominant grass species are concentrated at the origin.

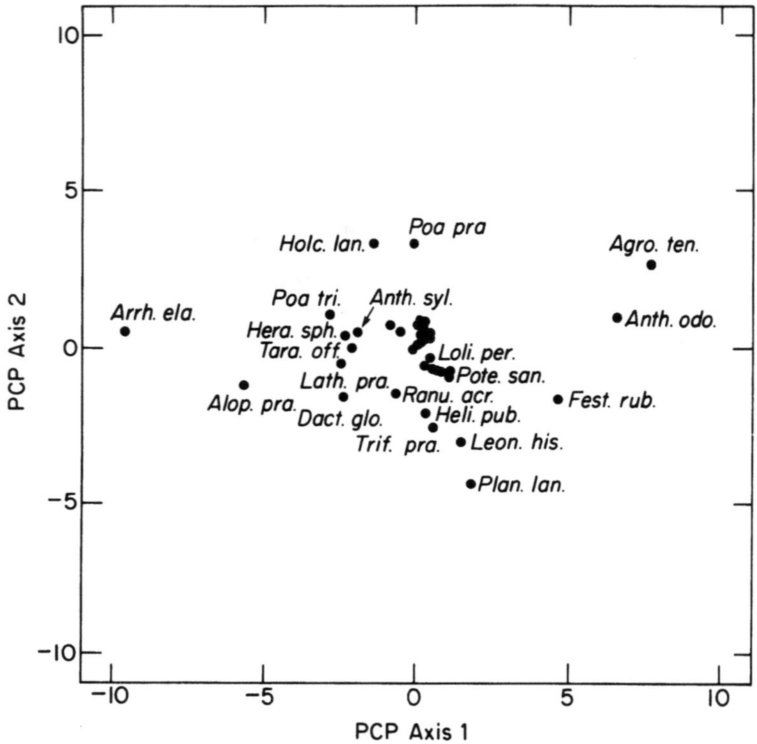

Fig. 3.5 Plot of scores for the Park Grass species on the first two principal components axes from an analysis of their relative abundances transformed by logarithms and doubly centred to have zero means. Named species occur in at least one field plot with an abundance of more than 5% of the plot's total dry matter yield.

tend to occur together in the same plot and thus show similar response to any major differences in environmental factors over the study area. In some cases the two-dimensional representation may be misleading – e.g. in over-emphasizing the proximity of *Agrostis* and *Anthoxanthum* (cf. Fig. 3.4 and Table 3.3) – but the primary ordinations of the species might generally be expected to be related to their environmental responses. However, in its present form the graphical representation does not allow the species ordination to be identified with any specific environmental gradient. It is thus interesting to see whether an ordination of the field plots based on species abundances is more readily interpretable, given that here additional information is available on treatments.

This field plot ordination is obtained by working with the transposed matrix $\mathbf{X}'$ (transformed to logarithms), where rows now represent plots and columns species. (Adjustment for a size component is now unnecessary since species abundances total 100% in all plots.) Figure 3.6 shows the PCP scores for the

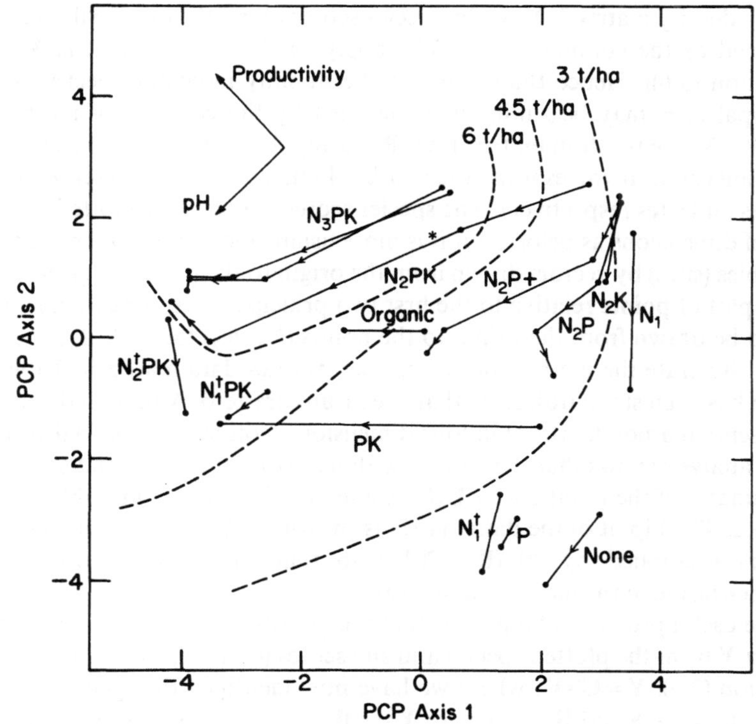

Fig. 3.6 Ordination of the Park Grass plots on the two principal components (logged data). Lines connect plots with the same fertilizer treatment but with amounts of lime, and hence soil pH, increasing in the direction of the arrows. Contours identify plots with similar levels of total dry matter yield. * indicates the true origin of the plots (see Section 3.5.5).

38 plots on the first two principal axes. The two-dimensional ordination of field plots is seen to span two environmental gradients of soil pH and plot productivity which run approximately at right angles to each other and at 45° to the two principal axes.

Clearly the species ordination (Fig. 3.5) would be much more informative if it could include some of the information seen in the field plot ordination. A method for combining the two will now be described.

3.2.3 Biplots

Suppose we start with a transformed species (units) by sites (variables) matrix $\mathbf{Y}$ $(n \times p)$ and add p dummy species to act as specific indicators for each of the p sites: i.e. the ith indicator species is given the value 1 at the ith site and 0 elsewhere. In matrix terms we can think of extending the $(n \times p)$ matrix $\mathbf{Y}$ by a

$(p \times p)$ identity matrix **I**. Now the species scores along the principal axes were obtained as the columns of $\mathbf{A}=\mathbf{YV}$, applying the rotation matrix **V** to **Y** (equation (3.6)). Hence the scores for the dummy indicator species on the principal axes may be similarly represented by the columns of the matrix $\mathbf{B}=\mathbf{IV}=\mathbf{V}$, the rotation matrix itself. Reducing the data matrix to a subspace of k dimensions involves using the first k columns of **A** and **B** to represent the species and sites respectively. The species scores may now be plotted for each pair of dimensions as before, but it is more usual, and useful, to represent the variables (sites) by vectors drawn from the origin to the positions given by $\mathbf{B}_k$. On a plot of points relative to the first two principal axes, for example, lines would be drawn from the origin to the points (b_{i1}, b_{i2}) for $i=1, 2, \ldots, p$.

We illustrate the method on the species-by-sites data matrix in Table 3.5 which has been standardized so that site means are zero. Although the matrix represents ten points (rows) in four dimensions (columns), principal components analysis shows that the data lie within a two-dimensional subspace; the coordinates of the points, $\mathbf{A}_2=\mathbf{U}_2\mathbf{S}_2$ are in fact the same as in Table 3.4 and Fig. 3.2. The biplot of the data matrix is shown in Fig. 3.7(a) with lengths of vectors multiplied by 20 ($\mathbf{B}_2=20\mathbf{V}_2$) so that the sites are on a scale commensurate with that for the species.

One useful property of biplots is that it is possible to approximate the data matrix **Y** from the plotted species and site scores (Gabriel, 1971, 1981). From equation (3.5), $\mathbf{Y}=\mathbf{USV}'$, where we have now identified the species and site scores as $\mathbf{A}=\mathbf{US}$ and $\mathbf{B}=\mathbf{V}$. Hence $\mathbf{Y}=\mathbf{AB}'$ or, in terms of the elements of **A** and **B**,

$$y_{ij}=a_{i1}b_{j1}+a_{i2}b_{j2}+\ldots+a_{ip}b_{jp}=\sum_{l=1}^{p} a_{il}b_{jl}. \tag{3.8}$$

Table 3.5 Artificial species-by-sites data set for biplot display.

	Sites			
Species	1	2	3	4
1	−6.68	−6.27	12.80	−3.06
2	−6.63	−2.15	11.65	0.90
3	−4.21	−0.99	7.31	0.93
4	−1.60	−4.57	3.86	−3.70
5	0.63	−4.38	0.07	−4.52
6	−0.57	2.80	0.22	2.96
7	2.48	−0.42	−4.04	−1.52
8	3.83	4.76	−7.64	2.88
9	5.56	6.17	−10.90	3.47
10	7.18	5.04	−13.31	1.65

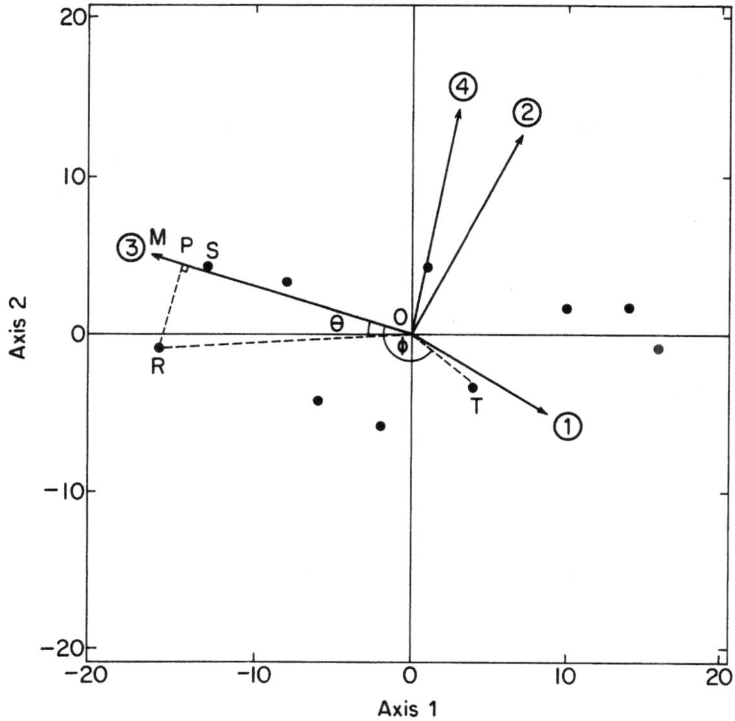

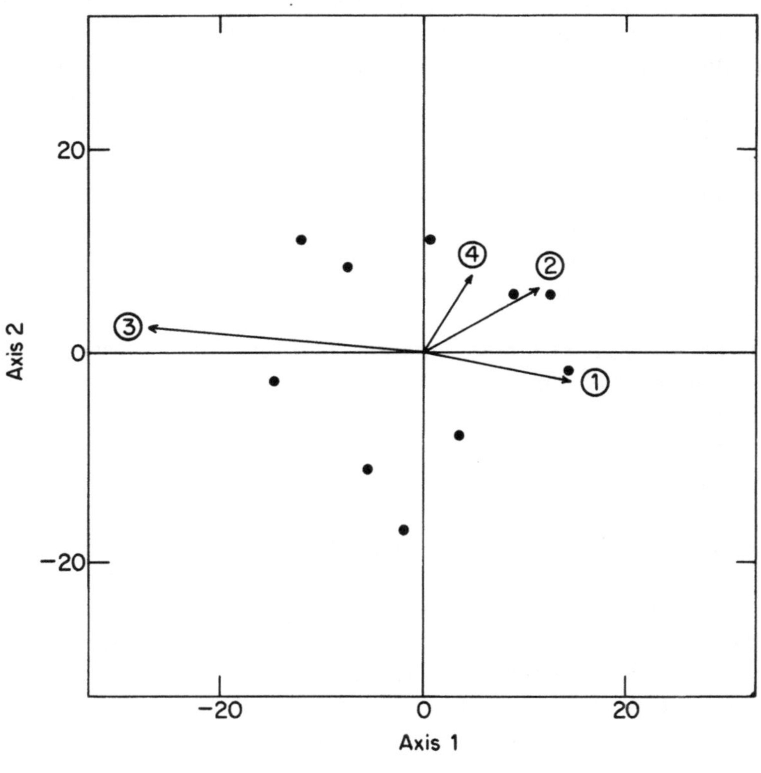

Fig. 3.7 Two forms of biplot for the species by sites data matrix of Table 3.5.
(a) Species PCP scores are plotted with the eigenvectors for the sites.
(b) An alternative representation where the length and direction of the site vectors indicate their standard deviations and correlations.

In itself this is not particularly useful as it merely confirms that the data can be recomposed from the analysis. However, if the data is reduced to a k-dimensional subspace, $\mathbf{A}_k$ and $\mathbf{B}_k$, it is possible to reconstruct an approximation to the data values, $\mathbf{Y}_k = \mathbf{A}_k\mathbf{B}'_k$, whose elements are given by

$$y_{ij}^{(k)} = \sum_{l=1}^{k} a_{il}b_{jl}. \tag{3.9}$$

For example, in the species-by-sites matrix of Table 3.5, the response of species (row) 1 in site (column) 3, $y_{13} = 12.80$, is approximated in two dimensions by

$$y_{13}^{(2)} = a_{11}b_{31} + a_{12}b_{32}. \tag{3.10}$$

A useful geometrical representation of these responses exists. In Fig. 3.7(a) the point for the first species $\mathrm{R} = (a_{11}, a_{12})$, while the point for the third site $\mathrm{M} = (b_{31}, b_{32})$. It can be shown (see e.g. Kempton, 1984) that the response in equation (3.10) may be derived from Fig. 3.7(a) as the inner product of OM and OR, viz

$$y_{13}^{(2)} = \mathrm{OM} \cdot \mathrm{OP} = \mathrm{OM} \cdot \mathrm{OR} \cdot \cos\theta,$$

where P is the projection of R onto OM. We illustrate this by obtaining approximate values for the quantities from the graph, and give the calculated exact values in parentheses for comparison: length OM = 17 (17.13) and length OP = 15 (14.94), or length OR = 16 (16.03) and $\cos\theta = \cos 20$ ($\cos 21.24$) = 0.94 (0.93), to give $y_{13}^{(2)} = \frac{17}{20} \times 15 = 12.75$ (12.80). (Note that we have divided OM by 20, since we use $\mathbf{B}_2 = 20\mathbf{V}_2$ rather than $\mathbf{V}_2$ in the figure.) In this example $y_{13}^{(2)} = y_{13}$ exactly, since the data matrix had been especially constructed to be represented exactly in two dimensions, but in general the adequacy of the approximation will depend on the goodness of fit of the two-dimensional subspace, as indicated by the relative magnitudes of the eigenvalues.

Let us now consider the representation of the set of species responses at one site, for example the values y_{i3} ($i = 1, 2, \ldots, 10$) of one complete column in the matrix of Table 3.5. Since the length of OM is a common factor of all the y_{i3} values, we can order these simply by the order of the projections of the species points onto the vector for the third site. Thus we can see that the species with largest response at the third site is species 1 (y_{13}), since the point P is the largest projection onto OM; in particular we note that y_{13} is greater than y_{23}, even though the point for the second species S is closer to M than R is. It is important to note the distinction between this type of graphical interpretation and the more usual distance interpretation. Note also that the sign of y_{ij} is determined by the angle subtended by the species on the site vector, and so the direction of the vector is important. For example, the point T for the seventh species subtends an angle $\phi = 165°$ on OM so that $\cos\phi = -0.96$, and $y_{73}^{(2)} = -4.0$.

Two other forms of biplot are commonly used: they differ in the way in which the singular values are allocated to **A** and **B**, i.e. the scaling given to the species and sites in the graph. However, since we still have $\mathbf{Y}_k = \mathbf{A}_k\mathbf{B}_k'$, the inner product formula can still be used to approximate **Y**. The biplot with $\mathbf{A} = \mathbf{US}^{1/2}$ and $\mathbf{B} = \mathbf{VS}^{1/2}$ is proposed by Bradu and Gabriel (1978) as a diagnostic tool to identify simple models which describe the data (Section 3.2.5). When the columns of the data matrix can be considered as variables, the biplot with $\mathbf{A} = \mathbf{U}$ and $\mathbf{B} = \mathbf{VS}$ may be informative. Figure 3.7(b) shows a biplot of this form for the previous data matrix, with the minor modification, $\mathbf{A} = 30\mathbf{U}$, to obtain commensurate scales for the row and column points. An additional interpretation of the site vectors is now available. From equation (3.5) we have

$$\mathbf{Y'Y} = (\mathbf{VSU'})(\mathbf{USV'}) = (\mathbf{VS})(\mathbf{SV'}) = \mathbf{BB'}$$

Hence the squared distance of the length of the jth site vector (the jth diagonal element of $\mathbf{BB'}$) is equivalent to the sum of squared responses for the jth site, which is proportional to the variance of responses, since sites have been standardized to have zero means. Also, the cosine of the angle between two vectors gives the correlation in species responses for the two sites. From Fig. 3.7(b) we see that site 3 shows greatest variability in species responses, while site 4 is the least variable. Sites 1 and 3 show a strong negative correlation in species response, whereas sites 1, 2 and 4 are positively correlated. These features can be confirmed from inspection of the data matrix (Table 3.5).

3.2.4 Example of a biplot

To show how effective the interpretation from a two-dimensional biplot may be in practice, Fig. 3.8 shows vectors, $\mathbf{B}_2 = \mathbf{V}_2$, representing six of the Park Grass field plots added to the two-dimensional plot of species scores, $\mathbf{A}_2 = \mathbf{U}_2\mathbf{S}_2$, given in Fig. 3.5 (the remaining 32 field plots have been omitted for the sake of clarity). The actual pattern of responses of the dominant species in these plots is summarized from Table 1.1 as follows:

Plots with N_2PK *treatment* (9a–d). On the unlimed plot (9d), *Anthoxanthum* is dominant (72% of total dry matter yield) with smaller amounts of *Agrostis* (15%) and *Holcus* (13%) present; on the recently limed plot (9c) there has been a large increase in yield with *Holcus* (44%) taking over as the most dominant species followed by *Poa pratensis* (16%), while *Anthoxanthum* and *Agrostis* have both been reduced to about 10% of total yield; on the continuously limed plots (9a and 9b), which have substantially higher pH, *Arrhenatherum* is dominant (52%), with *Alopecurus* (12%) and the legume *Lathyrus pratensis* (13%) both fairly abundant.

Unmanured plots (3a, 3d). In contrast, these plots have a more diverse flora with *Festuca* dominant, particularly on the unlimed plot, and *Agrostis*,

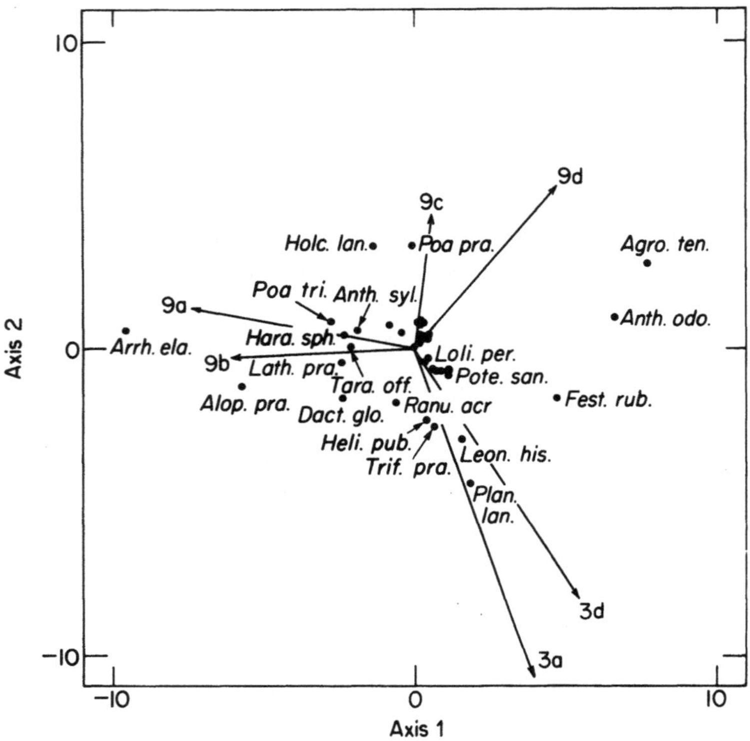

Fig. 3.8 Biplot for Park Grass showing the PCP scores for species (as in Fig. 3.5) and vectors for six field plots.

Anthoxanthum, *Helictotrichon*, *Leontodon*, *Plantago*, *Poterium* and *Trifolium repens* all contributing 5–10% of total dry matter yield.

An approximation to the species responses in each field plot is obtained from the biplot of Fig. 3.8 by projecting the species point onto the respective plot vector. While the values so derived show quantitative differences from the actual relative abundances in specific cases, there is a broad match with the overall pattern of responses described above.

In order to relate species and sites in the same biplot it is helpful to plot both on a similar scale. Frequently this requires that the coordinates of one set of points are multiplied by an appropriate constant open to choice, for example 20 and 30 as used in Figs 3.7 and 3.8 respectively.

3.2.5 Using biplots for model diagnosis

Bradu and Gabriel (1978) have proposed using biplots to suggest appropriate mathematical models for the values of a two-way table; see also Gabriel (1981).

In general we make little reference to modelling in this book, although as we shall see in Chapter 6, it is useful to know the type of graphical patterns produced by some of the simpler mathematical models (see also Fig. 2.18). In this context, Fig. 3.9 shows biplot displays for four different types of model. A graphical representation of the simple model for additive row and column effects, $y_{ij}=\alpha_i+\beta_j$, is displayed in Fig. 3.9(a): the points for rows and columns lie on two straight lines intersecting at 90°. When the lines do not intersect at 90°, as in Fig. 3.9(b), one can deduce that an interaction term is present: this may be expressed as a product of row and column terms, $y_{ij}=\alpha_i+\beta_j+\lambda\alpha_i\beta_j$. For the particular case when both lines pass through the origin, the simple proportional model $y_{ij}=\gamma_i\delta_j$ holds: such a model is appropriate for

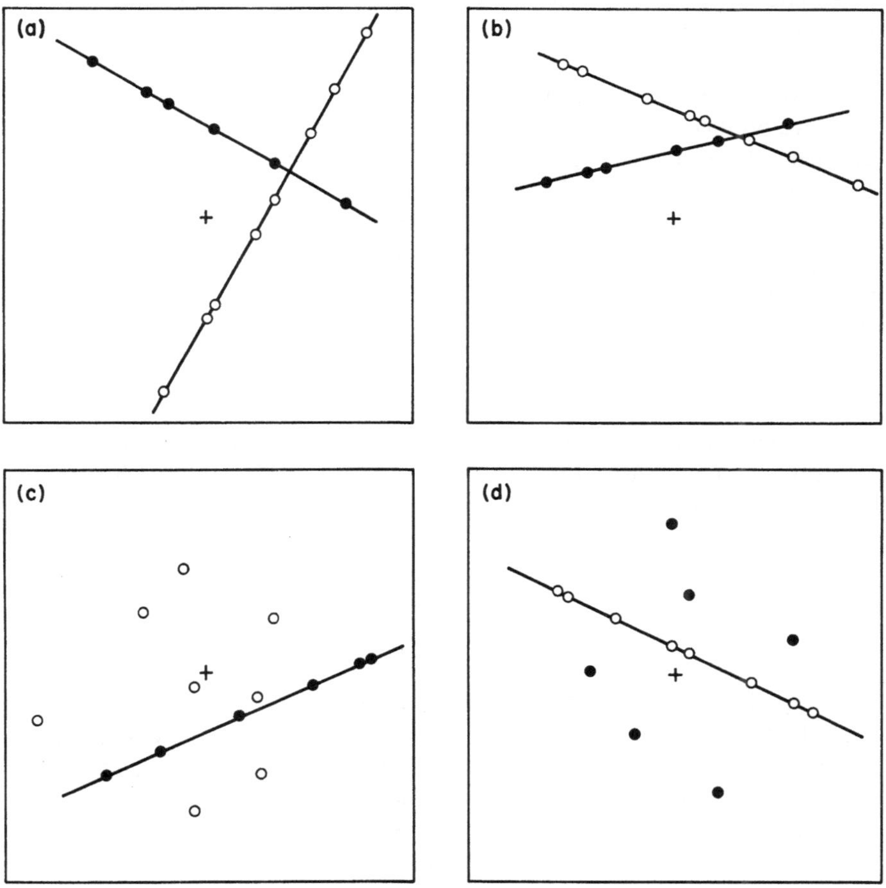

Fig. 3.9 Diagnostic biplots for selecting appropriate models for a two-way table. (a) When row (solid) and column (open) points lie on straight lines intersecting at right angles, an additive model will fit the data; while plots (b), (c) and (d) represent different forms of interaction between rows and columns. + indicates the origin.

contingency tables when row and column effects are independent. Finally, Figs 3.9(c) and (d) are indicative of the row and column regression models; these have more general interaction terms, and are of the form $y_{ij}=\alpha_i+\beta_j+\delta_j\alpha_i$ and $y_{ij}=\alpha_i+\beta_j+\gamma_i\beta_j$, respectively.

3.3 CORRESPONDENCE ANALYSIS

Correspondence analysis (CA) is now probably the most popular ordination method among ecologists. It is considered by some to be synonymous with reciprocal averaging (RA), although the French school that coined the term *l'analyse des correspondances* would maintain that a difference exists. Theoretically the methods are the same, apart possibly from scaling of axes, but correspondence analysis has been developed for wider practical application (Greenacre, 1984). We will first describe RA by considering an extension of direct gradient analysis.

As we pointed out in Section 3.1, direct gradient analysis takes a known gradient, of the sites for example, and uses this to construct a set of scores for the species. These scores themselves form a gradient (in that they can be ordered) and this species gradient may be used to obtain a new gradient of the sites; the new sites gradient can now be used to obtain a new species gradient, and so on. This iterative procedure is termed reciprocal averaging by Hill (1973).

3.3.1 Reciprocal averaging

Starting with a data matrix **X** of n species (rows) by p sites (columns), with elements x_{ij} representing the response of the ith species at the jth site, we obtain species scores a_i and site scores b_j by iteratively solving the sets of equations:

$$a_i=\rho^{-1}\sum_j x_{ij}b_j/r_i \qquad i=1,\ldots,n, \tag{3.11}$$

$$b_j=\rho^{-1}\sum_i x_{ij}a_i/c_j \qquad j=1,\ldots,p, \tag{3.12}$$

where $r_i=\Sigma_j x_{ij}$ and $c_j=\Sigma_i x_{ij}$ are the totals for the ith species and jth site, respectively, and the scaling parameter ρ is included to keep the scores within the same range from one iteration to another (cf. equations (3.1) and (3.2)).

Equations (3.11) and (3.12) are termed the transition formulae because they express each set of scores in terms of the other set. They always have the simple solution, $a_i=1$, $b_j=1$ and $\rho=1$, which is of no interest and is therefore removed from the data before applying the iterative procedure. The data are thus transformed to

$$y_{ij}=x_{ij}-x_{i.}x_{.j}/x_{..}, \tag{3.13}$$

where the second term on the right-hand side is that from the trivial solution.

The equations are now

$$a_i=\rho^{-1}\sum_j y_{ij}b_j/r_i, \tag{3.14}$$

$$b_j=\rho^{-1}\sum_i y_{ij}a_i/c_j. \tag{3.15}$$

To start the RA process we choose an arbitrary set of scores $b_j^{(0)}$ for the sites with range from 0 to 100, say; let $\rho=1$ and calculate species scores $a_i^{(1)}$ using equation (3.14). Then insert the species scores into equation (3.15) to obtain a revised set of site scores $b_j^{(1)}$. These site scores are then scaled to lie between 0 and 100 and the operations are repeated to obtain $a_i^{(2)}$ and $b_j^{(2)}$, and so on. Equations (3.14) and (3.15) possess a unique solution for the *a*'s and *b*'s, and the iterative procedure will converge to this solution whatever scores are used for starting values. Convergence is usually rapid but will depend on the size of the data matrix and the closeness of the starting values to the final solution. Hill (1973) gives a worked example.

Figure 3.10 shows, for the Park Grass data, the plot scores b_j for the first ten iterations starting from two initial gradients, plot pH and plot productivity. The single RA solution to which both converge is shown across the centre of the plot. Convergence is achieved more rapidly starting from the productivity gradient as this is closer to the final ordination.

3.3.2 The correspondence analysis approach

So far we have only displayed the data along a one-dimensional axis. The reciprocal averaging procedure can be easily extended to a second dimension by removing the first solution from the data, as we did with the trivial solution, and repeating to find an additional RA solution. This process can be repeated, giving an extra solution each time. However, rather than adopt this computational approach, we prefer to proceed via matrix algebra and find the full set of solutions; this is more in the theme of correspondence analysis.

Using matrix algebra, the transition formulae (3.11) and (3.12) are

$$\mathbf{a}=\rho^{-1}\mathbf{R}^{-1}\mathbf{X}\mathbf{b} \tag{3.16}$$

and

$$\mathbf{b}=\rho^{-1}\mathbf{C}^{-1}\mathbf{X}'\mathbf{a}, \tag{3.17}$$

where $\mathbf{R}$ and $\mathbf{C}$ are diagonal matrices with elements r_i and c_j representing species and site totals. Substituting from equation (3.16) into equation (3.17) we have

$$\mathbf{b}=\rho^{-2}\mathbf{C}^{-1}\mathbf{X}'\mathbf{R}^{-1}\mathbf{X}\mathbf{b} \tag{3.18}$$

and after some manipulation

$$\rho^2\mathbf{C}^{1/2}\mathbf{b}=(\mathbf{R}^{-1/2}\mathbf{X}\mathbf{C}^{-1/2})'(\mathbf{R}^{-1/2}\mathbf{X}\mathbf{C}^{-1/2})(\mathbf{C}^{1/2}\mathbf{b}),$$

where $\mathbf{R}^{1/2}$ and $\mathbf{C}^{1/2}$ are obtained from $\mathbf{R}$ and $\mathbf{C}$ by taking the square roots of

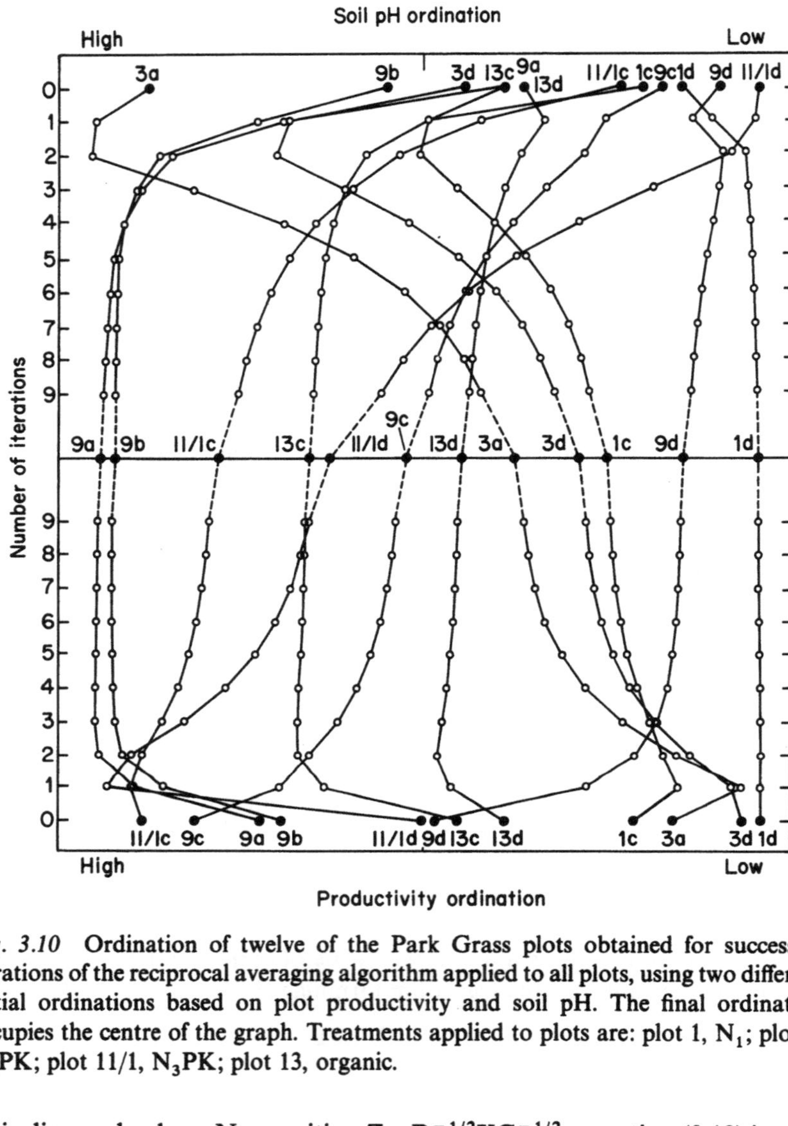

Fig. 3.10 Ordination of twelve of the Park Grass plots obtained for successive iterations of the reciprocal averaging algorithm applied to all plots, using two different initial ordinations based on plot productivity and soil pH. The final ordination occupies the centre of the graph. Treatments applied to plots are: plot 1, N_1; plot 9, N_2PK; plot 11/1, N_3PK; plot 13, organic.

their diagonal values. Now writing $\mathbf{Z}=\mathbf{R}^{-1/2}\mathbf{X}\mathbf{C}^{-1/2}$, equation (3.18) is seen to be merely the latent roots and vectors equation $\lambda\mathbf{v}=(\mathbf{Z}'\mathbf{Z})\mathbf{v}$, with $\lambda=\rho^2$ representing the latent roots and $\mathbf{v}=\mathbf{C}^{1/2}\mathbf{b}$ the latent vectors of the sum of squares and products matrix $\mathbf{Z}'\mathbf{Z}$.

Recalling that the spectral decomposition of $\mathbf{Z}'\mathbf{Z}=\mathbf{V}\mathbf{\Lambda}\mathbf{V}'$ is related to the singular value decomposition (SVD) $\mathbf{Z}=\mathbf{U}\mathbf{S}\mathbf{V}'$ with $\mathbf{\Lambda}=\mathbf{S}^2$, shows that the values of ρ for all the solutions are the singular values of $\mathbf{Z}=\mathbf{R}^{-1/2}\mathbf{X}\mathbf{C}^{-1/2}$ and the corresponding scores $\mathbf{b}$ are the columns of $\mathbf{B}=\mathbf{C}^{-1/2}\mathbf{V}$. Alternatively, we can substitute for $\mathbf{b}$ in the former RA equation and use a similar argument to that above to get $\mathbf{A}=\mathbf{R}^{-1/2}\mathbf{U}$, whose columns give the scores **a**, again for all the solutions.

Since we have derived **Z** from **X**, rather than from **Y**, the first column of **A**, and that of **B**, and the first singular value, will correspond to the trivial solution found by the RA process. Thus we need to consider **A** and **B** stripped of their first column. Now the scores for the first (non-trivial) solution are given in the first column of the two reduced matrices **A** and **B** and the graph of the first pair of solutions contains points (a_{i1}, a_{i2}) and (b_{j1}, b_{j2}).

3.3.3 Example of correspondence analysis

From equations (3.14) and (3.15), correspondence analysis is seen to provide an intrinsic scaling of the data by row and column means, so transformation of the data matrix prior to analysis is usually unnecessary. When applied to the untransformed species abundances on the Park Grass plots, the ordination of species and plots in the first two non-trivial dimensions is shown in Fig. 3.11. The corresponding singular values for this solution are $s_1=0.803$, $s_2=0.672$. The ordination of major species is similar to the ordination produced by a principal components analysis of their log-transformed abundances centred to have zero means (Fig. 3.5). However the scaling by species means, which is an intrinsic part of correspondence analysis, has given much greater weight to the rare species. The ordinations of field plots from the two analyses (Figs 3.6 and 3.11(b)) show greater similarities than those for species, probably because the original data were already scaled so that all plot (column) totals equalled 100%. Note that the ordination of field plots on axis 1 of Fig. 3.11(b) is precisely that given by the reciprocal averaging algorithm in Fig. 3.10.

3.3.4 *L'analyse des correspondances*

The French school has a somewhat different approach to correspondence analysis. The derivation is via the ideas of classical mechanics. Benzecri (1973) gives a very full description and Greenacre (1984) gives an account, in English, with examples. The method varies from that above only in the form of the matrices of scores $\mathbf{A}=\mathbf{R}^{-1/2}\mathbf{US}$ and $\mathbf{B}=\mathbf{C}^{-1/2}\mathbf{VS}$, so that the singular values are subsumed into the scores. See also Section 3.5.3 which links correspondence analysis with principal coordinates analysis.

The effect on the graphical representation is simply to compress the scales of the axes in proportion to their singular values. Thus, for the Park Grass ordination, the second axis would be compressed to a proportion $0.672/0.803=0.84$ of its span in Fig. 3.11, relative to the first axis.

3.3.5 Relation to original data matrix

It is common practice in correspondence analysis to plot points for both

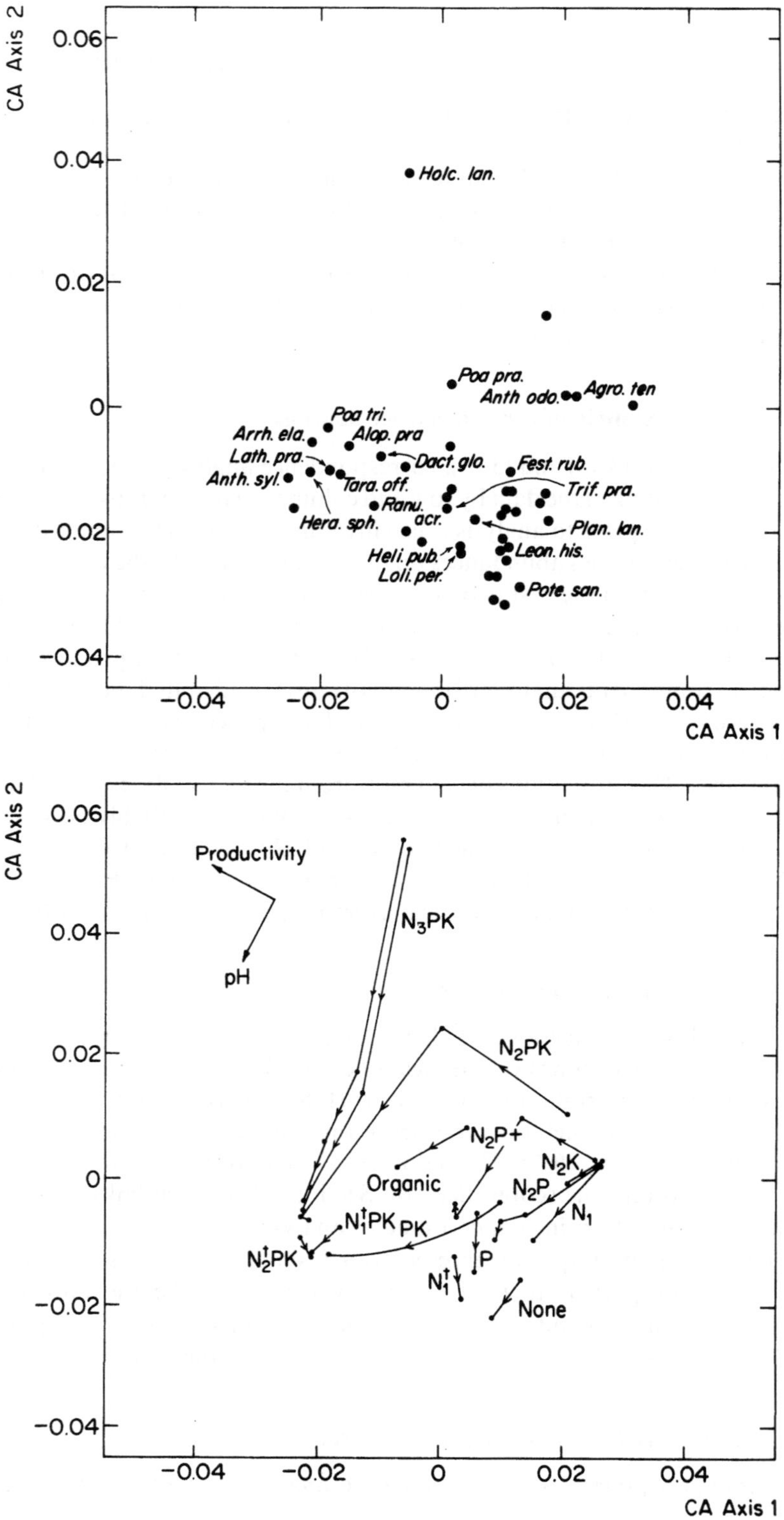

Fig. 3.11 Correspondence analysis for Park Grass data. Ordinations of (a) species and (b) field plots are shown separately, for clarity.

species and sites on the same graph. We have already shown (Section 3.2.3) that, for principal components analysis, approximations to the original data values can be found from such biplots. Clearly it would be useful if a similar result could be shown to hold for correspondence analysis plots.

Now from Section 3.3.2 we may write

$$\mathbf{R}^{-1}\mathbf{X}\mathbf{C}^{-1}=\mathbf{R}^{-1/2}\mathbf{Z}\mathbf{C}^{-1/2}=\mathbf{R}^{-1/2}\mathbf{U}\mathbf{S}\mathbf{V}'\mathbf{C}^{-1/2}=\mathbf{A}\mathbf{S}\mathbf{B}',$$

or equating the elements of the matrices on both sides

$$\frac{x_{ij}-r_ic_j}{r_ic_j}=\sum_{l>1} s_l a_{il} b_{jl}, \qquad (3.19)$$

where the second term in the numerator of the left-hand side comes from the trivial solution, (see Section 3.3.1), which is consequently omitted from the summation on the right-hand side. Dividing numerator and denominator on the left-hand side by the overall total abundance $\Sigma_i\Sigma_j x_{ij}$, the correspondence analysis solution is seen to relate to the difference between the data values, expressed as proportions of the grand total, and their expectations under the assumption of independence of species and sites. However, in contrast to equation (3.8) from the PCP formulation, the correspondence analysis solution includes the singular values with the product of species and site scores, which restricts direct inference about the data.

Suppose, however, that the scores on the individual axes are multiplied by the square roots of their respective singular values. Then the matrices of species and site scores are given by $\mathbf{A}=\mathbf{R}^{-1/2}\mathbf{U}\mathbf{S}^{1/2}$ and $\mathbf{B}=\mathbf{C}^{-1/2}\mathbf{V}\mathbf{S}^{1/2}$, and equation (3.19) becomes

$$\frac{x_{ij}-r_ic_j}{r_ic_j}=\sum_{l} a_{il} b_{jl}.$$

Assuming that the first two singular values (ignoring again the trivial solution) are relatively large compared to the remainder, this can be approximated by the first pair of axes:

$$\frac{x_{ij}-r_ic_j}{r_ic_j}\simeq a_{i1}b_{j1}+a_{i2}b_{j2}.$$

The expression on the right-hand side can again be interpreted geometrically as the inner product of species and site vectors OA_i, OB_j (cf. Section 3.2.3) and increases with distance of the points (a_{i1}, a_{i2}) and (b_{j1}, b_{j2}) from the origin, and also with the cosine of the angle between the vectors joining these points to the origin, O. Thus we might infer that x_{ij} is larger than expected under species/site independence if the relevant species and site points are close to each other but distant from the origin; conversely, two distant points on either side of the origin imply that x_{ij} is smaller than expected.

3.4 ORDINATION METHODS WHEN ROWS OR COLUMNS ARE GROUPED

Previous methods of ordination developed in this chapter make no allowance for any structure in the data matrix; although such structure may of course be subsequently incorporated into the graphical display. The two methods described here take into account any grouping of rows or columns of the data matrix. For example, in our context, the rows of an abundance matrix may represent species which can be grouped into families, while the columns may represent replicate samples from different sites.

A more traditional area of application for these techniques is in taxonomy, where the data matrix may represent observations on a number of individuals (rows) for a set of variables (e.g. morphometric measurements). If the individuals are grouped, say into species, *canonical variate analysis* (CVA) sets out to emphasize separations among species; it can therefore be used for discriminant analysis (Blackith and Reyment, 1971). This approach relies on assumptions being made about the distributional properties of the data which are rarely appropriate for the typical species-by-sites abundance matrices that form the basic consideration of this book. However, CVA provides an ordination of group means adjusted for variation within groups and it is as an ordination method that we shall be principally concerned with it here.

When the data matrix consists of observations for two sets of variables on the same set of units, for example sets of morphological and behavioural data for a number of individuals, *canonical correlation analysis* finds combinations of each set of variables so that these combinations, thought of in pairs (one for each set of variables), correlate as highly as possible. An alternative, and in our view more effective, solution when there are two or more sets of variables, is described in Chapter 4 under the title of Procrustes rotation.

3.4.1 Canonical variate analysis

This is also known as (linear) discriminant analysis, under which title it is fully described in Mardia, Kent and Bibby (1979); Seal (1966) discusses the method, under the title canonical analysis. The purpose of the method is to examine separations among a set of groups of units.

We assume that the data for n units (individuals) and p variates (morphometric measurements) are in the matrix $\mathbf{X}\,(n \times p)$; the n units are grouped into g classes (species) and correspondingly $\mathbf{X}$ is subdivided into g submatrices $\mathbf{X}_1, \mathbf{X}_2, \ldots, \mathbf{X}_g$. The data matrices are first transformed into $\mathbf{Y}$ and $\mathbf{Y}_1, \mathbf{Y}_2 \ldots, \mathbf{Y}_g$ by subtracting the column means of the respective data matrices; this corresponds with the normal standardization for PCP, but note that here the column means will differ from one submatrix to another. In PCP we obtained the sums of squares and products (SSP) matrix for the p variables as $\mathbf{Y}'\mathbf{Y}$: here we note that this is the total SSP matrix and write $\mathbf{T} = \mathbf{Y}'\mathbf{Y}$. Within each group one can also construct an SSP matrix, $\mathbf{W}_i=\mathbf{Y}_i'\mathbf{Y}_i$, and sum these to obtain the overall within-groups SSP matrix, $\mathbf{W}=\Sigma\mathbf{W}_i$. The between-group SSP may now be calculated as the difference of the total and within-group SSP, viz $\mathbf{B}=\mathbf{T}-\mathbf{W}$.

Canonical variate analysis seeks linear combinations of the p variates that have greatest between-group variation relative to their within-group variability. The first linear combination $\mathbf{v}_1$ is chosen to maximize the ratio of the between-group to within-group variation, $\mathbf{v}_1'\mathbf{B}\mathbf{v}_1/\mathbf{v}_1'\mathbf{W}\mathbf{v}_1$; the second linear combination maximizes the ratio of the variation subject to $\mathbf{v}_1'\mathbf{W}\mathbf{v}_2=0$, and so on. It can be shown (see e.g. Mardia, Kent and Bibby, 1979) that the required linear combinations are the eigenvectors of $\mathbf{W}^{-1/2}\mathbf{B}\mathbf{W}^{-1/2}$ while the corresponding eigenvalues denote the successive maxima of the ratio. Since a ratio is being maximized, it is immaterial how the $\mathbf{v}$ are scaled: the conventional form is so that $\mathbf{v}'\mathbf{W}\mathbf{v}=1$. It is worth noting that the number of linear combinations obtained from the spectral decomposition of $\mathbf{W}^{-1/2}\mathbf{B}\mathbf{W}^{-1/2}$ will be $\leqslant \min(p, g-1)$, where rank $(\mathbf{W}^{-1})=p$ and rank $(\mathbf{B})\leqslant g-1$, i.e. no more than $\min(p, g-1)$ of the eigenvalues of $\mathbf{W}^{-1/2}\mathbf{B}\mathbf{W}^{-1/2}$ will be non-zero.

As with PCP, the set of eigenvectors projects the transformed data matrix $\mathbf{Y}$ onto the multidimensional space that best separates the groups; this is termed the canonical variate space. However, it is more usual to construct only the smaller matrix $\mathbf{G}$ of group means and project these; here we have $g_{ij}=x_{.j}^{(i)}-x_{.j}$ where $x_{.j}^{(i)}$ and $x_{.j}$ are the means of the jth columns of $\mathbf{X}_i$ and $\mathbf{X}$, respectively. Now the constraint $\mathbf{v}'\mathbf{W}\mathbf{v}=1$ comes into play: since the within-group variation is equally represented in all directions in canonical variate space, circles may be drawn around each projected group mean to show the scatter of the units within the groups. If the original variates are normally distributed, a circle of radius $\sqrt{\chi^2_{2(P=0.95)}}=2.45$ should contain on average 95% of the units in the group.

3.4.2 Example of canonical variate analysis

Many good case studies may be found in the literature describing the use of canonical variate analysis to provide ordinations of species on the basis of morphological or behavioural characters (see e.g. Delany and Healy, 1964; Blackith and Reyment, 1971). The following example uses CVA as an ordination method. The data come from the Rothamsted Insect Survey (Taylor *et al.*, 1981) and were used in a study of the annual changes in distribution of moth species caught in light traps between 1969 and 1974 at 14 environmentally stable sites in the UK (Kempton, 1981). A total of 510 moth species were caught in that period but in this example we shall only consider 12 major species which are chosen, somewhat arbitrarily, because each was the single most abundant species at one of the sites in 1974 (one was the most abundant species at three sites). Table 3.6 shows, for each species, the mean

Table 3.6 Mean abundance (log numbers/year) of 12 moth species at 14 sites in the UK, and their year-to-year variability within sites from 1969 to 1974.

Site code, number, environmental category† and name	*Operophtera brumata*	*Orthosia gothica*	*Lithina chlorosata*	*Agrostis exclamationis*	*Diarsia rubi*	*Spilosoma lutea*	*Luperina testacea*	*Lithosia lurideola*	*Omphaloscelis lunosa*	*Oporinia dilutata*	*Lygris populata*	*Perizoma affinitata*
A 22 **W** Rothamsted, Geescroft	5.3	4.4	0.9	3.9	3.3	4.3	0.9	1.9	1.3	7.0	0.0	0.1
B 16 **W** Stratfield Mortimer	4.5	3.9	2.2	3.8	1.9	3.9	3.9	3.1	2.6	4.6	0.0	0.0
C 46 **W** Alice Holt	2.9	3.7	3.1	3.8	2.1	3.7	1.9	3.7	5.1	3.7	0.0	0.8
D 78 **W** Ringwood	1.3	3.8	4.9	3.1	2.7	4.7	2.9	3.2	2.2	0.7	0.0	0.0
E 92 **P** Nettlecombe Court	2.5	5.1	2.4	3.5	4.1	4.5	2.4	6.0	3.7	2.7	0.7	2.3
F 67 **F** Slapton Ley	0.7	3.6	0.7	5.3	5.0	4.5	4.2	4.3	2.3	0.9	0.0	1.9
G 126 **F** Aberystwyth	1.4	2.9	2.2	3.6	4.5	3.8	3.5	1.5	2.0	0.6	0.4	3.4
H 94 **P** Monks Wood	0.2	2.3	0.0	4.5	4.7	3.1	2.3	2.3	3.6	1.7	0.0	0.0
I 88 **F** Broom's Barn	1.0	2.1	0.2	3.5	4.1	2.3	4.3	3.4	2.2	1.0	0.1	0.2
J 127 **W** Kielder	3.3	5.9	0.2	0.7	3.4	0.1	0.4	0.0	0.0	3.4	4.4	0.0
K 29 **P** Rannoch	2.5	5.4	0.4	0.0	4.7	0.0	0.0	0.0	0.0	2.6	6.0	0.0
L 49 **W** Fort Augustus	2.6	5.7	5.3	0.7	3.9	0.0	0.1	1.5	0.0	0.3	2.4	0.0
M 58 **W** Elgin	1.9	4.9	0.9	2.4	3.2	0.0	2.4	0.3	0.0	1.2	1.8	0.2
N 57 **W** Ardross	3.8	5.9	0.5	0.3	4.9	0.0	0.0	0.0	0.0	4.7	4.9	1.5
Total within-site sum of squares	46.7	12.5	36.5	55.0	36.5	14.9	28.3	24.4	34.4	94.1	21.1	20.2

† Environmental categories are: **W**, woodland; **P**, parkland; **F**, farmland. (From Taylor *et al*., 1981.)

abundance at each site over the six years and the year-to-year variability in abundance pooled over all sites. Although a log transformation has been adopted to standardize the abundances, there is still an eight-fold difference in variability among species. A principal components ordination of sites based on the means of Table 3.6 would give equal weight to all species. However, species which have an erratic appearance at a site are unlikely to be reliable indicators for the habitat and should be given a lower weight than species with more consistent abundances when constructing an ordination. If correlations between species across years are small, as in this example, CVA effectively carries out a weighted PCP, with species' weights inversely proportional to their within-site variability in abundance. The two-dimensional ordination from CVA is shown in Fig. 3.12 and accounts for 76.4% of the total variation among sites (compared with 72.7% for PCP). The first axis clearly separates northern from southern sites, while the second axis appears to separate predominantly woodland sites in southern England from other English sites. A more detailed analysis of these data, with all species included, will be given in Chapter 4.

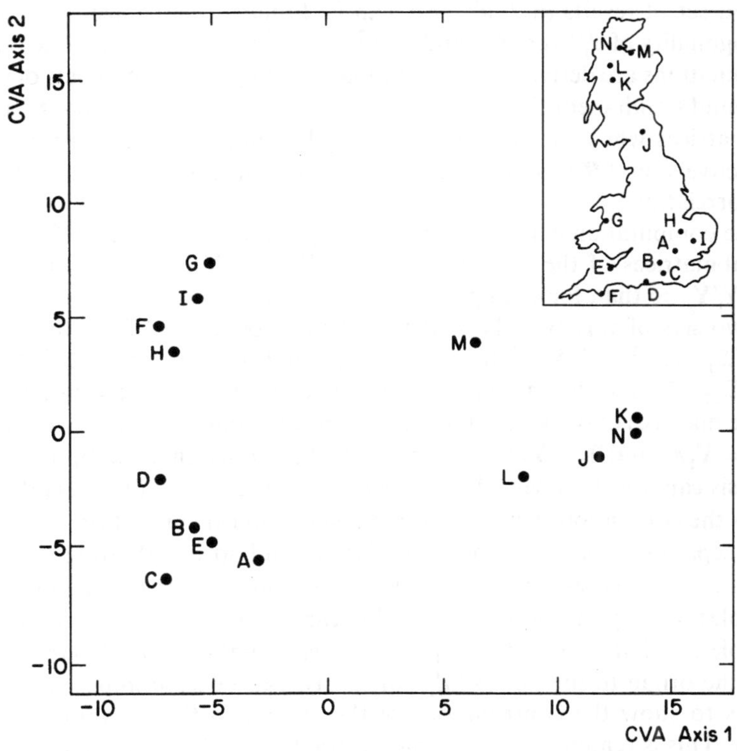

Fig. 3.12 Ordination of moth samples from 14 locations in UK (see inset) using canonical variate analysis. For site codes see Table 3.6.

3.4.3 Canonical correlation analysis

This method can be used when the set of variables is divided into two subsets. Most commonly, we may have records of species abundances and also information on environmental factors for a number of sites; Gittins (1979), in an extensive review of the method, considers several sets of data of this type. Canonical correlation analysis sets out to find linear combinations of the two sets of variables for the sites such that the linear combinations have maximal correlation. The procedure is repeated to obtain an independent second pair of linear combinations that have maximal correlation, and so on. Thus we might look for combinations of the species abundances that best correlate with combinations of the environmental variables so that species distributions may be predicted from environmental information on the site.

For convenience we assume that the data matrix $\mathbf{X}$ can be partitioned as $(\mathbf{X}_1 | \mathbf{X}_2)$ where $\mathbf{X}_1$ and $\mathbf{X}_2$ are the two sets of variables. Subtracting column means as usual gives $(\mathbf{Y}_1 | \mathbf{Y}_2)$. The method finds linear combinations, $\mathbf{b}^{(1)} = \mathbf{Y}_1 \mathbf{a}^{(1)}$ and $\mathbf{b}^{(2)} = \mathbf{Y}_2 \mathbf{a}^{(2)}$, to maximize the correlation r between $\mathbf{b}^{(1)}$ and $\mathbf{b}^{(2)}$. This gives the first set of canonical correlation results $(\mathbf{a}_1^{(1)}, \mathbf{a}_1^{(2)}, r_1)$. A second set of results $(\mathbf{a}_2^{(1)}, \mathbf{a}_2^{(2)}, r_2)$, can be found to maximize r_2 subject to orthogonality of $\mathbf{b}_1^{(1)}$ and $\mathbf{b}_2^{(1)}$, and of $\mathbf{b}_1^{(2)}$ and $\mathbf{b}_2^{(2)}$; third and subsequent linear combinations are derived in a similar manner. In practice these sets of results are found simultaneously, to give the complete set $(\mathbf{A}_1, \mathbf{A}_2, \mathbf{R})$ where $\mathbf{A}_1$ and $\mathbf{A}_2$ are matrices whose jth columns are the jth canonical vectors $\mathbf{a}_j^{(1)}$ and $\mathbf{a}_j^{(2)}$, respectively, and $\mathbf{R}$ is a diagonal matrix with elements r_j $(j = 1, 2, \ldots)$ giving the correlations.

One computational approach to find $(\mathbf{A}_1, \mathbf{A}_2, \mathbf{R})$ (Banfield, 1978) operates on submatrices of the SSP matrix $\mathbf{\Sigma} = \mathbf{Y}'\mathbf{Y} = (\mathbf{Y}_1 | \mathbf{Y}_2)'(\mathbf{Y}_1 | \mathbf{Y}_2)$. First define $\mathbf{\Sigma}_{ij} = \mathbf{Y}_i'\mathbf{Y}_j$, so that, for example, $\mathbf{\Sigma}_{12}$ is the matrix of cross-products between the two sets of variables. Now the results can be derived from the SVD of $\mathbf{\Sigma}_{11}^{-1/2}\mathbf{\Sigma}_{12}\mathbf{\Sigma}_{22}^{-1/2} = \mathbf{USV}'$. The linear combinations are $\mathbf{A}_1 = \mathbf{\Sigma}_{11}^{-1/2}\mathbf{U}$ and $\mathbf{A}_2 = \mathbf{\Sigma}_{22}^{-1/2}\mathbf{V}$, and the canonical correlations are the singular values $\mathbf{S}$.

The matrices of site scores for the two sets of variables are easily calculated as $\mathbf{B}_1 = \mathbf{Y}_1\mathbf{A}_1$ and $\mathbf{B}_2 = \mathbf{Y}_2\mathbf{A}_2$. Let us denote the (i, j)th element of $\mathbf{B}_k$ by $b_{ij}^{(k)}$. The analysis can now be assessed graphically in various ways. Plotting $(b_{i1}^{(1)}, b_{i1}^{(2)})$ shows the correlation structure for the first canonical correlation; this can be done separately for the second correlation, and so on. Plotting $(b_{i1}^{(1)}, b_{i2}^{(1)})$ shows the site scores for the first two linear combinations of the first set of variables, e.g. species abundances; this can be done for the second set of variables, and also for different pairs of linear combinations. Vectors drawn from the origin to the points $(a_{i1}^{(1)}, a_{i2}^{(1)})$, say, can be superimposed on these graphs to show the contributions of the species to the relevant canonical vector. This is reminiscent of the biplot (Section 3.2.3) and similar inferences can be drawn.

3.4.4 Example of canonical correlation analysis

In Section 3.2.2 we saw that the variation in the relative abundances of species on the Park Grass plots could be largely described by differences in the acidity and overall productivity of the plots, brought about by the different fertilizer treatments. We now investigate the interrelationship between the species frequencies and the actual treatments themselves. Table 3.7 gives the major treatments applied to the limed (a) and unlimed (d) plots. However, rather than using the 44 species frequencies themselves to characterize the plots

Table 3.7 Treatments applied to Park Grass (plots a and d only) and scores on first four axes from a principal components analysis of flora on all plots (Section 3.2.2). For treatment codes see Table 1.1.

	Treatments					*Plant species*			
Plot code	N	N†	P	K	*Lime*	*Axis 1*	*Axis 2*	*Axis 3*	*Axis 4*
1d	1	0	0	0	0	3.54	1.77	−1.73	0.85
3a	0	0	0	0	1	2.11	−4.06	0.02	−1.70
3d	0	0	0	0	0	2.99	−2.94	−0.11	−0.46
4a	2	0	1	0	1	1.91	0.11	2.46	2.09
4d	2	0	1	0	0	3.31	2.26	−2.62	0.28
7a	0	0	1	1	1	−3.33	−1.45	−2.12	0.36
7d	0	0	1	1	0	2.00	−1.49	−0.11	−0.06
8a	0	0	1	0	1	1.36	−3.47	−0.07	−1.00
8d	0	0	1	0	0	1.62	−3.02	0.29	−1.94
9a	2	0	1	1	1	−4.16	0.59	−0.27	0.19
9d	2	0	1	1	0	2.81	2.57	−1.30	−1.54
10a	2	0	1	0	1	0.09	−0.28	1.66	1.82
10d	2	0	1	0	0	3.33	2.28	−2.51	0.33
11/1a	3	0	1	1	1	−3.86	1.11	0.86	−0.92
11/1d	3	0	1	1	0	0.52	2.42	0.52	−3.49
11/2a	3	0	1	1	1	−3.87	0.98	0.42	−0.50
11/2d	3	0	1	1	0	0.36	2.52	0.72	−3.46
14a	0	2	1	1	1	−3.91	−1.27	−1.70	1.96
14d	0	2	1	1	0	−4.19	0.30	−1.37	1.18
16a	0	1	1	1	1	−3.33	−1.43	−1.71	1.49
16d	0	1	1	1	0	−2.54	−0.89	−1.21	0.12
17a	0	1	0	0	1	1.02	−3.88	0.11	−1.40
17d	0	1	0	0	0	1.35	−2.60	−0.68	−0.60
18d	2	0	0	1	0	3.31	2.38	−2.45	0.38
Variance explained by principal components (%)						38.1	16.6	10.5	9.1

(Table 1.1) we use the plot scores from the principal components analysis (Fig. 3.6); since we found in Section 3.2.2 that the first four principal components account for 74% of the total variation in species log abundances over the plots, while the fifth component accounts for less than 5% of the total, these four components provide a succinct description of the overall pattern of the 44 species. Furthermore, in this instance, replacing the species by the principal components scores considerably aids interpretation of the analysis. These treatment and plant values now provide the two sets of variables for a canonical correlation analysis.

Table 3.8 shows the four canonical vectors for both sets of plant and treatment variables and the corresponding canonical correlations. Corresponding vectors in the two sets have maximal correlation subject to orthogonality with the previous vectors in the same set. For plants, the first and second canonical vectors are closely related to the second and first principal axes, respectively. For treatments, the first two canonical vectors may be approximated by 4N + P + 2K − 3L and 3N† + 2P + 4K + 4L, respectively, and these are very closely correlated with the plant vectors ($r_1 = 0.98$, $r_2 = 0.90$). Hence the ordination of plots along the first principal axis of Fig. 3.6 separates those plots receiving N† (as sodium nitrate), P, K and lime from those without these treatments, while the second axis separates those plots receiving N (as ammonium sulphate), P and K, but not lime, from other plots. Summing these two vectors, we have a diagonal vector indicating plots with increasing N, P and K, and hence increasing productivity; and by subtracting them we have a cross-diagonal vector for increasing lime, and hence soil pH. This confirms the earlier interpretation.

The correlations between the third and fourth canonical vectors are too small to be worth further investigation.

Table 3.8 Canonical correlation analysis of plant and treatment variates in Table 3.7. All canonical vectors are standardized to have unit sum of squares.

Canonical vector		I	II	III	IV
Correlation		0.981	0.900	0.592	0.253
		Linear combinations			
Plants	*Axis 1*	−0.10	0.97	0.08	−0.15
	Axis 2	0.88	0.01	0.03	−0.22
	Axis 3	0.36	0.04	0.89	0.29
	Axis 4	−0.31	−0.22	0.44	−0.92
Treatments	N	0.72	0.01	0.19	−0.16
	N†	0.13	−0.42	0.06	−0.67
	P	0.20	−0.32	0.29	−0.28
	K	0.38	−0.57	−0.72	0.59
	Lime	−0.53	−0.63	0.60	0.32

3.5 PRINCIPAL COORDINATES ANALYSIS

The method of principal coordinates analysis (Gower, 1966a) has various synonyms; it is commonly called classical scaling, particularly by users of the gamut of techniques known collectively as multidimensional scaling. We shall use the mnemonic PCO hereafter, partly because it will be familiar to some readers and partly to avoid confusion with principal components analysis (PCP).

Previous methods in this chapter operate on a matrix of n units classified by p variables. In contrast, the input to PCO is a symmetric $(n \times n)$ matrix **A** with values a_{ij} that represent the associations among the n units. The symmetry condition implies that the association between the ith and jth units is the same as that between the jth and ith. (In Chapter 6 we consider methods of analysis when this symmetry condition does not hold.) As we observed in Chapter 1, the association matrix may be derived from a units-by-variables data matrix, by calculating, for example, a measure of association in the spatial distribution of pairs of species over a number of sites; alternatively, the original observations may themselves be made in the form of an association matrix. It is assumed that as the association measure a_{ij} increases, so does the proximity or similarity of the units, while their distance or dissimilarity decreases. Frequently **A** will be a similarity matrix with $a_{ii}=1, 0 \leqslant a_{ij} \leqslant 1$, but there is no requirement for associations to be simple similarity indices. However, if distances δ_{ij} are to be used as input they should initially be transformed as $a_{ij}=-\frac{1}{2}\delta_{ij}^2$.

3.5.1 Mathematical derivation

PCO gives a configuration of n points in up to $n-1$ dimensions such that the squared distance between points i and j is given by:

$$d_{ij}=a_{ii}+a_{jj}-2a_{ij}. \tag{3.20}$$

The reason for choosing this function of the elements of **A** becomes clear below.

Let us represent the configuration of points in an r-dimensional subspace by the matrix **Z** $(n \times r)$, where $r \leqslant n-1$. Now the squared distance between the ith and jth points, $(z_{i1}, z_{i2}, \ldots, z_{ir})$, $(z_{j1}, z_{j2}, \ldots, z_{jr})$, is

$$d_{ij}=\sum_{l=1}^{r}(z_{il}-z_{jl})^2, \tag{3.21}$$

and if the squared term is expanded we have

$$d_{ij}=\Sigma z_{il}^2+\Sigma z_{jl}^2-2\Sigma z_{il}z_{jl}. \tag{3.22}$$

Comparing equations (3.20) and (3.22), it is sufficient to find a matrix **Z** such that

$$a_{ij}=\sum_{k=1}^{r} z_{ik}z_{jk},$$

or, equivalently, $\mathbf{A}=\mathbf{ZZ}'$.

Since $\mathbf{A}$ is symmetric we can find its spectral decomposition: $\mathbf{A}=\mathbf{\Gamma\Lambda\Gamma}'$. For the moment we shall assume that all the eigenvalues are non-negative so that we can form the diagonal matrix $\mathbf{\Lambda}^{1/2}$ with values $\lambda_i^{1/2}$. If we then define the matrix of points as $\mathbf{Z}=\mathbf{\Gamma\Lambda}^{1/2}$, it follows immediately that $\mathbf{ZZ}'=\mathbf{\Gamma\Lambda}^{1/2}\mathbf{\Lambda}^{1/2}\mathbf{\Gamma}'=\mathbf{A}$, as required.

The SSP matrix of $\mathbf{Z}$ is $\mathbf{Z}'\mathbf{Z}=\mathbf{\Lambda}^{1/2}\mathbf{\Gamma}'\mathbf{\Gamma\Lambda}^{1/2}=\mathbf{\Lambda}$, since $\mathbf{\Gamma}$ is orthogonal ($\mathbf{\Gamma}'\mathbf{\Gamma}=\mathbf{I}$). Hence the columns of $\mathbf{Z}$ are mutually orthogonal and the variation explained by each dimension (column of $\mathbf{Z}$) is given by the appropriate eigenvalue λ of $\mathbf{A}$. Thus for any arbitrary column of $\mathbf{Z}$, $\mathbf{z}=(z_1, z_2, \ldots, z_n)'$, with mean $\bar{z}$,

$$\lambda=\mathbf{z}'\mathbf{z}=\Sigma z_i^2=n\bar{z}^2+\Sigma(z_i-\bar{z})^2.$$

Since we are mainly interested in the distances or variation among the n points, we would wish to remove variation due to the means $\bar{z}$ by setting the column sums of $\mathbf{Z}$ to zero. This can be achieved if the association matrix $\mathbf{A}$ is first double-centred by subtracting row and column means and adding the grand mean:

$$a_{ij}=a_{ij}-a_{i.}-a_{.j}+a_{..}$$

Note that, from the form of equation (3.20), this double-centring will have no effect on the squared distances d_{ij}, but it will affect the spectral decomposition; in particular, at least one eigenvalue will be zero. In practice $\mathbf{A}$ is always double-centred before the spectral decomposition is found. The diagonal values of $\mathbf{A}$, after double-centring, are in fact the squared distances of the points from their centroid, i.e. the origin; they may be useful for interpretation (Section 3.5.2).

To describe the $n(n-1)/2$ squared distances d_{ij} between the n objects exactly, it will usually be necessary to represent them in the full $n-1$ dimensions. The sum of squared distances between the points is then

$$\sum_{i,j} d_{ij}=\sum_{l=1}^{n-1} \lambda_l.$$

However, as with principal components (Section 3.2.1), the effectiveness of the method lies in the fact that the dominant patterns in the data are reflected in the first few dimensions (columns of $\mathbf{Z}$). As before, the total variation explained by the first k dimensions is

$$t_k=\sum_{l=1}^{k} \lambda_l$$

and when k is small, association among the objects may be displayed by plotting one dimension against another in the usual way.

3.5.2 Example of principal coordinates analysis

Yarranton (1966) describes a method of sampling vegetation which gives a direct measure of local association among species. This is based on point sampling, where both the species hit by the sampling point and the species touching the first species nearest to that point are recorded. In effect, Yarranton argues, the sampling point is removed to the line of contact between the two species so they have a reciprocal relationship. Where the species hit by the sampling point does not touch another plant, the sample is recorded as 'no contact'.

Yarranton (1966) applies the method to sampling mosses growing among rocks at Steps Bridge, Devon. The number of contacts among the more abundant species is given in Table 3.9. We use these contacts to measure the spatial association among pairs of species and so provide an ordination of species using principal coordinates analysis.

The observations on inter-specific contacts are strongly dependent on the individual abundances of the species and so require some standardization before analysis. For each species, slightly over half the contacts are intra-specific. The chosen measure of association between species i and j standardizes the number of inter-specific contacts n_{ij} by the number of intra-specific contacts n_{ii} and n_{jj}:

$$a_{ij} = \log[(n_{ij}+1)/2\sqrt{(n_{ii}n_{jj})}].$$

For large samples, this measure takes the value 0 when inter-specific contacts are as likely as intra-specific, i.e. when two species are distributed independently and at random. For these data, all a_{ij} were negative, indicating that plants of the same species are highly aggregated.

The eigenvalues for the association matrix **A** were, in decreasing order of size:

23.4, 16.5, 15.6, 11.3, 10.3, 9.2, 8.4, 6.7, 5.7, 4.5, 4.3, 4.0, 2.9, 2.4, 2.0, 1.6, 1.1, 1.0, 0.6, 0.3, 0.2, 0, −1.0, −1.8, −2.1, −2.7, −2.9, −3.7.

We note that there is one zero eigenvalue, as expected, but, additionally, six negative values. This causes difficulties in interpretation; for example, the variation among the species explained by the first k dimensions increases with k for $k<22$, but then decreases. We shall discuss negative eigenvalues further in Section 3.5.6. For the present we observe that there are three dominant positive eigenvalues and show the species plotted on the first two principal axes in Fig. 3.13. This ordination shows species grouped according to their more usual habitats. The two axes provide a good separation of the habitat groups and suggest an environmental gradient of increasing shade and moisture from the top left-hand corner to the bottom right corner of the plot. The third axis further separates habitat groups A, B and C.

Since the first two eigenvalues are not exceptionally dominant, the distances

Table 3.9 Number of inter- and intra-specific contacts among rock mosses. (Data from Yarranton, 1966, using only species with more than 250 plants sampled.)

1 *Andreaea rothii*	1404													
2 *Bryum alpinum*	113	386												
3 *Bryum capillare*	3	1	152											
4 *Cephalozia* sp.	.	.	1	297										
5 *Cynodontium bruntonii*	9	8	.	5	537									
6 *Dicranella heteromalla*	.	.	3	106	25	431								
7 *Dicranoweissia cirrata*	1	.	2	.	.	7	233							
8 *Dicranum majus*	.	.	1	.	.	.	2	141						
9 *Dicranum scoparium*	3	.	4	55	15	61	10	.	559					
10 *Diphyscium foliosum*	.	.	.	.	28	2	.	.	7	245				
11 *Diplophyllum albicans*	20	.	.	.	11	2	37	.	7	18	433			
12 *Frullania tamarisci*	.	.	13	.	.	.	4	.	3	.	.	147		
13 *Grimmia apocarpa*	20	.	3	.	5	.	1	.	24	4	.	2	286	
14 *Grimmia montana*	53	1	1	.	1	.	.	.	.	.	.	.	4	214

15 *Grimmia trichophylla*	18	.	8	.	.	.	.	1	8	3	.	.	8	15	223													
16 *Hypnum cupressiforme*	23	1	66	31	54	61	66	19	196	11	20	57	120	4	74	1794												
17 *Isopterygium elegans*	.	.	.	7	30	12	.	.	1	3	.	.	.	.	.	15	211											
18 *Isothecium myosuroides*	1	.	3	.	18	9	23	68	40	31	22	11	11	.	11	181	62	1083										
19 *Metzgeria* spp	5	.	.	1	38	19	10	.	24	1	26	4	4	.	4	55	11	106	773									
20 *Microlejeunea ulicina*	3	.	1	.	13	5	2	.	1	.	.	6	.	.	3	72	3	31	27	223								
21 *Plagiothecium denticulatum*	5	.	4	36	21	49	3	.	22	3	15	.	7	.	9	65	16	140	188	9	777							
22 *Pohlia nutans*	11	21	9	25	17	18	.	3	2	9	3	1	.	.	1	3	1	.	.	.	2	298						
23 *Polytrichum formosum*	.	.	.	.	.	1	3	20	7	11	4	.	.	.	.	24	.	91	13	.	12	.	194					
24 *Polytrichum piliferum*	362	136	13	.	9	.	.	22	6	2	.	.	17	53	15	78	1	2	.	.	.	117	.	1243				
25 *Rhabdoweissia fugax*	20	4	.	9	51	16	.	.	7	40	33	.	1	.	3	28	13	6	16	5	27	14	.	11	469			
26 *Rhacomitrium aquaticum*	197	61	.	.	8	1	.	.	.	7	31	.	4	.	12	32	9	11	24	3	36	1	1	27	35	781		
27 *Rhacomitrium heterostichum*	334	11	3	.	42	4	17	1	11	2	101	15	9	15	3	96	.	63	91	3	66	8	4	230	17	46	1366	
28 *Scapania compacta*	63	2	5	.	30	1	6	.	3	13	32	2	2	.	9	28	1	4	1	.	5	8	.	4	30	144	62	509

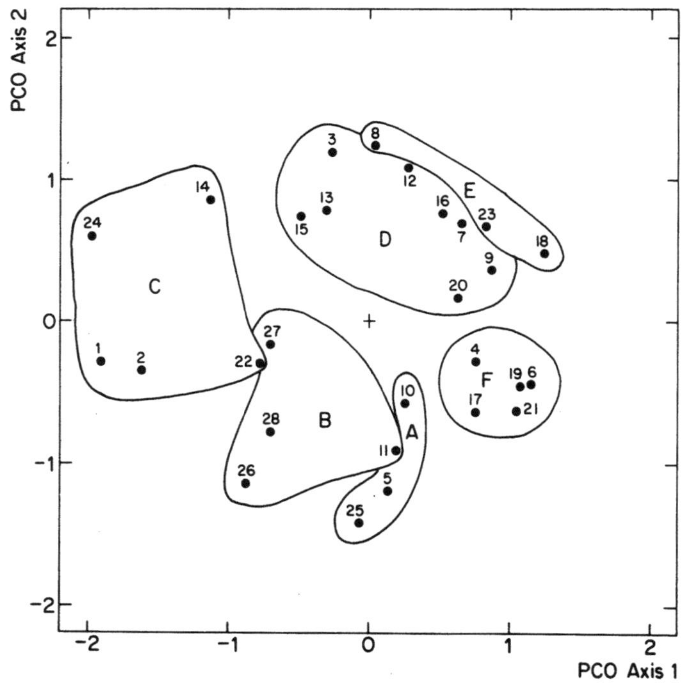

Fig. 3.13 Principal coordinates analysis of rock mosses based on the number of inter-specific contacts (Table 3.9). The species are grouped according to habitat type (Yarranton, 1966): A, crevice species; B and C, mosaic elements of exposed leached rock surfaces, B occurring in slightly more shaded places; D, species on unhumified dry rock surfaces; F, species of humified substrates; E, common species of rocky woodland ground flora not allocated to D or F.

among pairs of species coordinates on the two-dimensional ordination give only a rough approximation to the actual distances derived from the contact matrix of Table 3.9. For example, Table 3.9 shows that species 16, *Hypnum cupressiforme*, occurs relatively frequently with most other species, except species 1, 2, 14 and 22 from habitat group C: as a consequence it would be expected to lie close to the centroid of the species points. In fact it is the closest point, by some margin, to the centroid in the full set of dimensions; although in Fig. 3.13 eight other species lie closer to the origin than species 16. In contrast, species 2 and 14 lie furthest from the centroid in the full analysis and this is well represented in Fig. 3.13.

3.5.3 Correspondence between PCO and other methods

One of the most powerful aspects of PCO lies in the generality of the association matrix **A**. In ecology, naturally occurring data are less likely to be in the form of association matrices than they are in other areas, psychology for example; but an association matrix may always be derived from a rows and columns data matrix, as we saw in Section 1.4. If the original data are in a matrix $\mathbf{X}$ ($n \times p$) of, say, species by sites, $\mathbf{A}$ can be derived from $\mathbf{X}$ in any number of ways; here we consider particular methods of derivation which result in PCO solutions that mimic other methods of analysis already described.

The matrix $\mathbf{A}$ is typically derived from a matrix $\mathbf{D}$ of squared distances, $\mathbf{A} = -\mathbf{D}/2$, and different metrics may be used to form $\mathbf{D}$ from $\mathbf{X}$. The Euclidean metric gives the squared distance

$$d_{ij} = \sum_{k=1}^{p} (x_{ik} - x_{jk})^2 \tag{3.23}$$

and the PCO solution from $\mathbf{A} = -\mathbf{D}/2$ will be identical to that from principal components analysis (PCP) of $\mathbf{X}$. Note that here all the eigenvalues will be non-negative and at least $n-p$ will be zero.

Table 3.10 gives the calculated distances among the six most abundant grass species based on their relative abundances on the Park Grass plots. Distances among all species pairs were of similar magnitude and large compared with other less abundant species on the plots. A PCO based on the full distance matrix for all 46 species produces the PCP plots of Figs 3.3 and 3.4. The six most abundant species were clearly separated on the five axes while the less abundant species were all concentrated close to the origin, in accord with their relative inter-specific distances. Note also that *Alopecurus*, which is associated with somewhat smaller inter-specific distances in Table 3.10, particularly if these are adjusted for species means, took a more central position than the other five major species in Fig. 3.4. The logarithmic transformation of the abundances applied in Section 3.2.2 reduced the distances among the major species relative to the others and produced the more balanced ordination of Fig. 3.5.

In contrast, if only presence/absence records are used to provide an ordination of the species, the PCO is based on a similarity matrix such as that of Table 1.5 which uses the Czekanowski measure. As shown in Table 1.4, the six more abundant species appear in most of the plots and all the similarities

Table 3.10 Euclidean distances among six major grass species on Park Grass plots.

Holcus	—					
Agrostis	206	—				
Anthoxanthum	194	144	—			
Festuca	191	164	160	—		
Alopecurus	162	164	149	136	—	
Arrhenatherum	204	221	208	197	115	—

among them lie between 75 and 100%, while the less common species such as *Lolium*, *Leontodon* and *Poterium* have low similarity with all other species and each other. Hence a PCO ordination will tend to group the more abundant species at the centre, and separate the less common species on successive principal axes. This is clearly unsatisfactory and lends support to Greig-Smith's (1983) view that crude presence/absence information is not adequate for comparing local stands of vegetation.

Correspondence analysis can also be presented as a special case of principal coordinates analysis, by first standardizing the data matrix for rows and columns. The squared distances among the species, i.e. the rows of **X**, are given by the chi-squared metric,

$$d_{ij} = \sum_{k=1}^{p} \frac{(x_{ik}/r_i - x_{jk}/r_j)^2}{c_k} \tag{3.24}$$

where r_i and c_k are the totals of the ith row and kth column of **X**, respectively. Note that $d_{ij}=0$ only if the ith and jth rows have the same proportions. The PCO scores for the species can be superimposed on the same plot as the PCO scores for the sites, i.e. columns of **X**, which are derived from the corresponding metric,

$$d_{ij} = \sum_{k=1}^{n} \frac{(x_{ki}/c_i - x_{kj}/c_j)^2}{r_k}. \tag{3.25}$$

The inter-species and inter-site distances calculated from the complete correspondence analysis (i.e. the full set of dimensions) will be the same as those used by PCO, from equations (3.24) and (3.25). However, the approximate representation in (say) two dimensions will differ slightly. This is because the criteria used by the two methods in finding the best approximation are different. Note that the method of correspondence analysis referred to here is that associated with Benzecri (1973) and the French school (Section 3.3.4): an approximation to the reciprocal averaging plot is obtained by dividing the scores on each PCO axis by the square root of the corresponding eigenvalue.

Gower (1966b) also describes an approach whereby PCO can be used for a form of canonical variate analysis (CVA). This is achieved by working with the Mahalanobis distances between the group means: if the mean of the ith group is $\bar{x}_i$ and the pooled within-group variance–covariance matrix is **W**, then the Mahalanobis (squared) distance (Section 1.4.3) between the ith and jth group means is

$$d_{ij} = (\bar{\mathbf{x}}_i - \bar{\mathbf{x}}_j)\mathbf{W}^{-1}(\bar{\mathbf{x}}_i - \bar{\mathbf{x}}_j)'.$$

In particular, if the correlations among variables are negligible,

$$d_{ij} = \sum_{k=1}^{n} \frac{(\bar{x}_{ik} - \bar{x}_{jk})^2}{s_k^2},$$

where s_k^2 is the pooled within-group variance of the kth variable. The two sets of results, from CVA directly or via PCO using Mahalanobis distances, will not in general be identical unless all the groups are of equal size, but it is unlikely that any differences will be large enough to affect interpretation.

3.5.4 Linking close neighbours on PCO plot

A two-dimensional representation of inter-unit distances on a PCO plot may give a misleading impression of the actual proximity of units. A pair of units which appear close together in the first two dimensions may be far apart in higher dimensions. This is particularly likely when the first two principal coordinates fail to describe a large percentage of the total variation among the units. In these cases information from a two-dimensional plot can be increased by linking units whose actual distance apart is less than some critical value chosen by the user. Alternatively, units may be connected by their minimum spanning tree (Gower and Ross, 1969): this is the network of links having minimum total length such that every unit is linked in the tree and there are no closed loops – a simple illustration is given in Fig. 3.14 and a fuller description in Chapter 5.

3.5.5 Interpreting principal components analyses for site diversity

So far our discussion of distances has only concerned inter-site (or inter-species) distances or their distances from their common centroid. With principal coordinates analysis (PCO) the location of the origin is arbitrary and, as shown in the mathematical development, is most conventionally located at the centroid. However, with principal components analysis (PCP) a true origin exists, although in the usual analysis it is lost by the intrinsic centring of equation (3.3). Without this centring the distances from the sites to the true origin may be interpreted in terms of their species diversity (ter Braak, 1983). Suppose the elements of **X** represent the set of relative abundances x_{ij} of species j at site i, such that $\Sigma_j x_{ij} = 1$. Then the Euclidean squared distance from site i to the true origin is

$$d_i = \sum_j x_{ij}^2, \tag{3.26}$$

or, if the proportions are first transformed by logarithms as in Fig. 3.6,

$$d_i = \sum_j (\log x_{ij})^2. \tag{3.27}$$

Equations (3.26) and (3.27) define intrinsic diversity measures in the sense of Patil and Taillie (1982). Equation (3.26) is Simpson's inverse measure so that, with an uncentred PCP plot, sites with high diversity will be closer to the origin than those with low diversity. The reverse interpretation will apply when the log proportions are used.

In species-centred PCP the origin of the coordinate system is shifted from the true origin to the centroid of the sites. Thus to obtain approximate site diversities from the PCP plot we need to project the true origin onto this plot. (Because the true origin does not contribute to the centred analysis, this projection may lead to poor interpretations of the diversities; however, it will often be adequate where a general impression is all that is required.) Now the vector of species means is $(x_{.1}, x_{.2}, \ldots, x_{.n})$ and hence the position of the true origin along the lth principal axis is

$$z_l = -\sum_j v_{jl} x_{.j},$$

where v_{jl} is the jth element of the lth eigenvector from the uncentred PCP of the sites-by-species matrix **X**. The true origin of the Park Grass plots is shown on Fig. 3.6 for the PCP of logged proportions. This shows that the plots with high nitrogen, applied as ammonium sulphate, had a much lower species diversity than those without ammonium sulphate. The highest diversity occurred in the unmanured plot in the bottom right of Fig. 3.6. The diversity of ammonium sulphate plots generally increased with liming, indicating that low diversity was associated with low pH. As ter Braak (1983) notes, the addition of species points on the Figure, following the biplot method of Section 3.2.3, would help to indicate which species were contributing to the high diversity of a particular field plot.

3.5.6 Negative eigenvalues from principal coordinates analysis

We assumed in the derivation of PCO that all eigenvalues from the analysis of the association matrix **A** are non-negative. When **A** is derived from a Euclidean distance matrix, as described in Section 1.4.5, this assumption always holds. For example, **A** might have been derived from a rows and columns matrix using a distance measure that has the Euclidean property. However, negative eigenvalues are not uncommon when working with general association matrices; see e.g. Section 3.5.2. Dimensions (columns of **Z**) corresponding to negative eigenvalues are imaginary and cannot be plotted in the usual way.

Negative roots occur when the PCO model, i.e. the representation of associations by inter-point distances, cannot satisfactorily resolve all the data; the situation is best demonstrated by example. Suppose that we have three species, R, S and T, and the associations between them give rise to inter-point distances of 1, 1 and 4. Clearly it is not possible to arrange three points geometrically to satisfy these distances; in fact, the distances are non-metric as well as non-Euclidean (Section 1.4.5). The matrix of squared distances **D** is

R	0		
S	1	0	
T	1	16	0
	R	S	T

from which the PCO analysis gives latent roots (8, 0, −2). The coordinates of the three points in the first, and only real, dimension are 0, 2 and −2; this gives fitted squared distances

R	0		
S	4	0	
T	4	16	0

The squared distance ST has been obtained exactly but the others are too large. An exact fit to the squared distances would require an extra dimension in which the points for S and T were coincident and the point for R was a squared distance of -3 $(=1-4)$ from them; i.e. the distance of the point for R from the coincident points for S and T would have to be $\sqrt{-3}$, which is impossible in a real representation.

In practice negative roots are unlikely to occur as simple inconsistencies among three points, as above, but will generally arise from the interrelationships among many points. Sibson (1979) has investigated the effect of perturbing the distances corresponding to a PCO solution with no negative roots; such perturbations are typical of the sort of (statistical) error that might occur in ecological data. When these perturbed distances were analysed using PCO, a common result was a series of relatively small latent roots, some positive and some negative, that were distinct from the roots corresponding to the original solution; additionally the small positive roots tended to cancel out the small negative roots. This suggests that, in practice, when relatively small negative roots occur (as in Section 3.5.2) they can be ignored, together with an equivalent total of small positive roots.

However, large negative eigenvalues may seriously distort an analysis, particularly as they may go unnoticed when only the principal dimensions corresponding to the first few eigenvalues are displayed. Care must be taken over expressing the eigenvalues as percentages of the total variation since, when negative eigenvalues are present, the sum of the positive eigenvalues will exceed the total; consequently the amount of variation explained by the chosen number of principal dimensions will be overestimated, and may indeed be greater than 100%.

When large negative eigenvalues are present it is important to realize that the ordination is unsatisfactory in some respects. It may be advisable to investigate these further; for example, a comparison of the actual squared distances and those from a low-dimensional PCO solution may highlight particular groups of interrelationships that are the cause of the inconsistencies. Such a comparison is made most easily by plotting the fitted squared distances against those from the data.

3.6 THE HORSESHOE EFFECT

When analysing samples from diverse habitats along a single environmental gradient, it is often observed that an ordination on the two principal axes does not produce a linear sequence of sites or species. The curvilinearity may show itself as a mild arching effect, in which case the ordination is still clearly discernible on the primary axis, although possibly somewhat compressed at either end; however, when habitat diversity is large, the ordination may take the shape of a horseshoe with end-points involuted (Fig. 3.14(a)); hence the name of this phenomenon. In such extreme cases the ordination along the first axis will be partially obscured and, even with only moderate curvilinearity a secondary ecological gradient which would otherwise be shown up on the second axis may be hidden in a higher dimension. Sensitivity to this arching effect thus becomes an important criterion in choosing between different ordination methods.

3.6.1 Genesis

At first sight the dependence observed between the two axes is unexpected since the axes were chosen to be orthogonal. However, this constraint of orthogonality is not sufficient to guarantee the axes' independence, as is shown by considering the artificial data matrix in Table 3.11. This shows a simple one-dimensional gradient of sites based on presence–absence of 16 species. As is often observed, the number of species is greatest at sites in the centre of the range and tails off towards either extreme.

An unstandardized principal components analysis (Fig. 3.14(a)) gives a two-dimensional ordination with a typical horseshoe shape, where the ordering of sites on the first axis is lost. This distortion is easily explained by deriving the relevant distance matrix for PCP (Table 3.11(a)); in this case the squared distance between two sites is equal to the number of species which occur at one site or the other, but not at both. However, since sites at both ends of the gradient have fewer total species than sites in the centre of the gradient, the calculated distances between the two ends is less than expected, producing the observed involution.

The distortion effect can be reduced by using a more appropriate distance matrix. If, for example, we calculate a weighted Euclidean distance for each pair of sites so that species which occur at both sites are given double weight, while species missing from both sites have zero weight – the complement of Czekanowski's similarity coefficient for binary data (Section 1.4.1) – then we obtain the distance matrix of Table 3.11(b) in which sites with no species in common now have zero similarity and are a constant distance apart. The involution of sites on the first axis (Fig. 3.14(b)) is now almost removed, but sites at either end of the gradient are still very compressed. Correspondence analysis uses a form of distance matrix (Table 3.11(c)) which is standardized for both species and site totals (equation (3.24)), and has been shown by Gauch, Whittaker and Wentworth (1977) to produce a less distorted ordination than unstandardized PCP. In this example, the double standardization is shown to be very effective in removing the involution of end-points and producing an ordination of approximately equally spaced sites on the first axis (Fig. 3.14(c)).

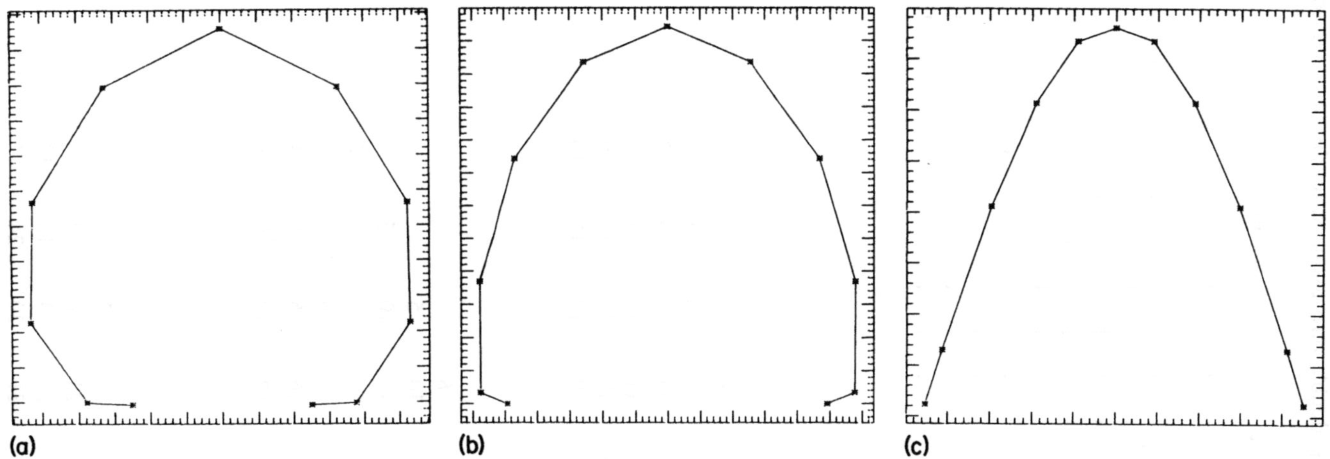

Fig. 3.14 Site scores on the first two dimensions of a principal coordinates analysis for the data of Table 3.11: (a) Euclidean distances including zero matches; (b) Euclidean distances excluding zero matches; (c) distances derived from correspondence analysis. The minimum spanning tree (Section 3.5.4) indicates a linear gradient of sites, but this is partially obscured in the ordination on the first axis for (a) and (b).

Table 3.11 Artificial presence–absence matrix for 16 species at 11 sites with derived inter-site distance matrix based on: (a) Euclidean metric (PCP distances); (b) weighted Euclidean metric ignoring double zero matches; (c) distances for correspondence analysis.

Species

	1	1	1	.	.	.	.	.	.	.	.	.	.	.	.	.
	1	1	1	1	.	.	.	.	.	.	.	.	.	.	.	.
	.	1	1	1	1	1	.	.	.	.	.	.	.	.	.	.
	.	.	1	1	1	1	1	1	.	.	.	.	.	.	.	.
	.	.	.	1	1	1	1	1	1	1	.	.	.	.	.	.
Sites	.	.	.	.	1	1	1	1	1	1	1	1	.	.	.	.
	.	.	.	.	.	.	1	1	1	1	1	1	1	.	.	.
	.	.	.	.	.	.	.	.	1	1	1	1	1	1	.	.
	.	.	.	.	.	.	.	.	.	.	1	1	1	1	1	.
	.	.	.	.	.	.	.	.	.	.	.	.	1	1	1	1
	.	.	.	.	.	.	.	.	.	.	.	.	.	1	1	1

Inter-site distances:

(a)	1									
	4	3								
	7	6	3							
	10	9	6	3						
	11	12	9	6	3					
	10	11	12	9	6	3				
	9	10	11	12	9	6	3			
	8	9	10	11	12	9	6	3		
	7	8	9	10	11	12	9	6	3	
	6	7	8	9	10	11	10	7	4	1
(b)	4									
	11	8								
	14	12	7							
	16	14	11	6						
	16	16	13	10	5					
	16	16	16	13	10	5				
	16	16	16	16	13	10	6			
	16	16	16	16	16	13	11	7		
	16	16	16	16	16	16	14	12	8	
	16	16	16	16	16	16	16	14	11	4
(c)	15									
	31	23								
	37	29	17							
	40	32	21	13						
	39	34	24	18	12					
	40	34	30	23	18	12				
	40	35	31	29	23	18	13			
	42	37	33	31	30	24	21	17		
	45	41	37	35	34	34	32	29	23	
	49	45	42	40	40	39	40	37	31	15

3.6.2 Detrending

A single dominant gradient describing the samples will usually be detected from the ordination diagram despite the presence of curvilinearity, particularly if near neighbours are connected as in Fig. 3.14. However, when the pattern of abundances is more complex or contains a large noise component, the presence of curvilinearity can then obscure a gradient or suggest a secondary gradient where none exists. Two methods for reducing this curvilinearity, i.e. detrending, are described below. The first of these is not in general use and appears to be restricted to presence–absence data. The second seems to us to be rather arbitrary and the precise details of the method are hidden in a computer program.

Williamson (1978) recognized that, at least for presence–absence data, the arching effect with PCO is largely caused by the presence of large numbers of pairs of sites with zero similarity (no species in common) which are all defined to be a constant distance apart. He argued that zero similarity should be regarded as indicating an unknown distance and proposed a 'step-across' method for estimating these distances using intermediate sites. Applying PCO to this amended distance matrix resulted in ordinations which were free from major distortion.

An alternative approach, developed for correspondence analysis by Hill and Gauch (1980), has been adopted more readily by ecologists because of the availability of a computer program (Hill, 1979a). The method, termed detrended correspondence analysis, consists of simple trend removal in successive axes with optional rescaling of axes to remove compression of points at either end. An example follows.

The species abundance matrix given in Table 3.2 for field plots from a park-meadow in south Sweden shows some similarity with the artificial data matrix of Table 3.11. Species and sites are here ordered according to their position along the first correspondence analysis axis which represents an environmental gradient of increasing shading and soil moisture. The second and third CA axes (Fig. 3.15(a), (b)) are clearly quadratic and cubic functions of the first, but the fourth axis (Fig. 3.15(c)) appears to identify a secondary gradient of plots with similar light and moisture scores but increasing soil pH and nitrogen (cf. Persson, 1981). Detrended correspondence analysis without rescaling (Fig. 3.16) successfully removes the curvilinearity in the primary gradient: axis 2 now isolates plot 13, as in axis 4 of the previous analysis.

3.7 NON-METRIC ORDINATION

Methods such as principal coordinates analysis operate with a set of inter-unit squared distances, either observed directly or derived from other data, and attempt to produce an ordination of the units, in a small number of dimensions, in which these distances are well represented. Non-metric

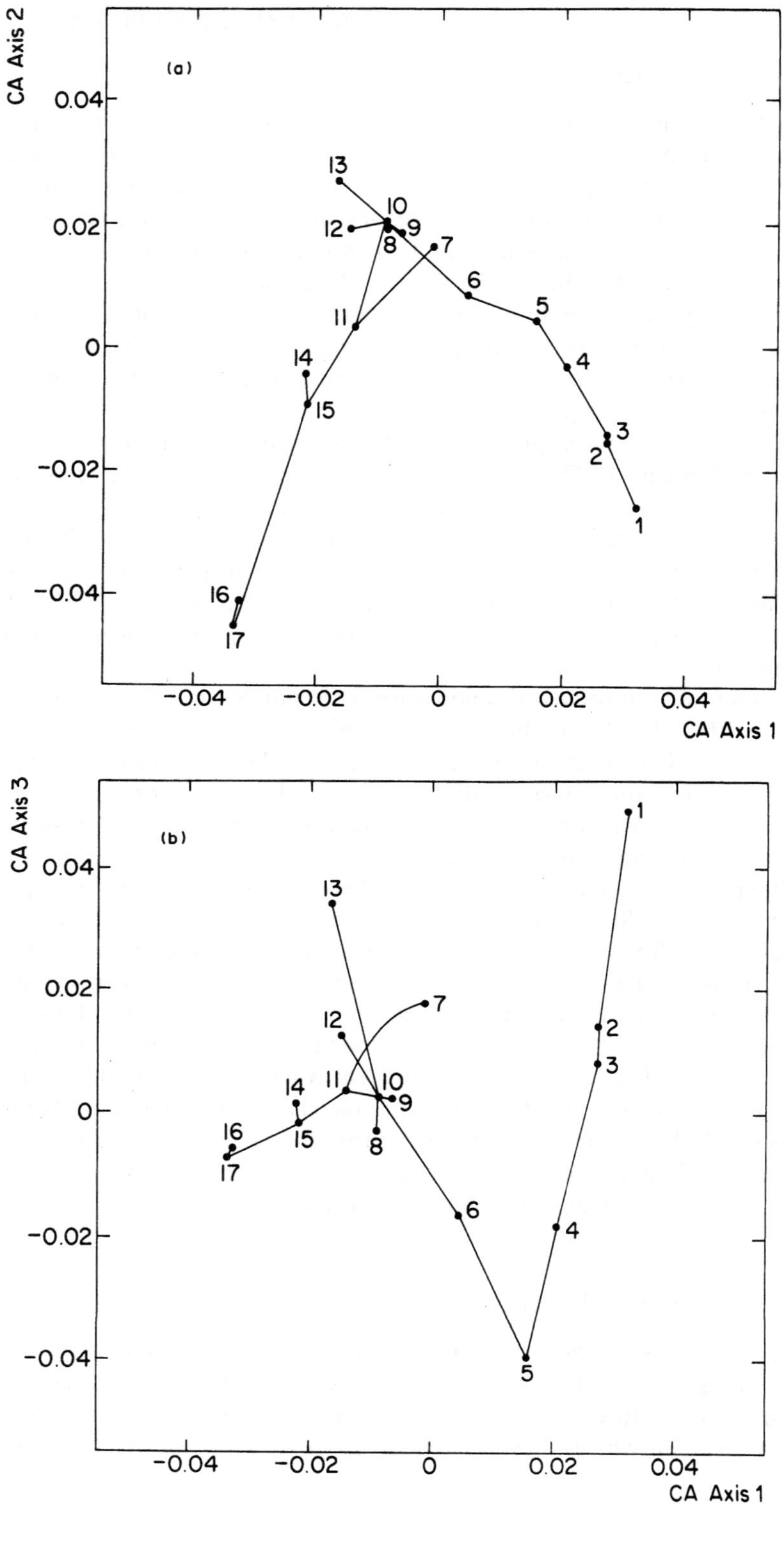

CA Axis 2
(a)
0.04
0.02
0
-0.02
-0.04
-0.04
-0.02
0
0.02
0.04
CA Axis 1
1
2
3
4
5
6
7
8
9
10
11
12
13
14
15
16
17
CA Axis 3
(b)
0.04
0.02
0
-0.02
-0.04
-0.04
-0.02
0
0.02
0.04
CA Axis 1

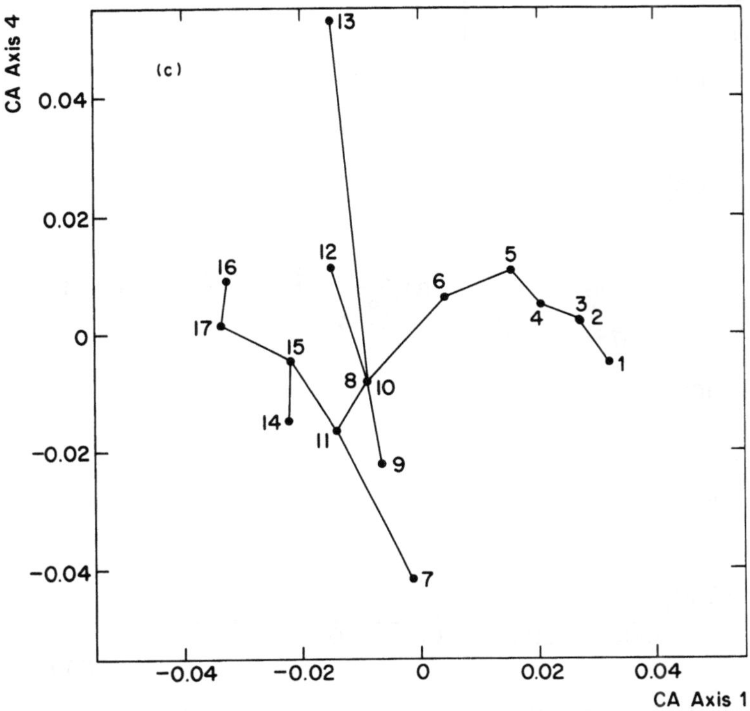

Fig. 3.15 Ordination of field plots from a park-meadow in Stenervd, Sweden, based on the first four dimensions of a correspondence analysis of species abundance scores (Table 3.2). Plots are connected by their minimum spanning tree using the χ^2 distances of equation (3.24).

methods also produce ordinations from distances; but rather than use the actual distances, only their ranked order is considered: hence the alternative name of ordinal scaling. The methods are commonly used under this name in the social sciences where measurement scales are often arbitrary and have limited intrinsic meaning. In ecology, it has been argued that non-metric ordinations are more robust to aberrant values (for example a species with an exceptionally high abundance at a site in one year), and hence are likely to be more consistent in repeated samples, say from year to year.

The non-metric solution for a given number of dimensions is given by the ordination which minimizes the STRESS function, usually calculated as

$$\text{STRESS} = \sqrt{\left\{ \sum_i \sum_{j<i} (\hat{d}_{ij} - \text{f}(d_{ij}))^2 \Big/ \sum_i \sum_{j<i} \hat{d}_{ij}^2 \right\}}. \tag{3.28}$$

In equation (3.28), $\hat{d}_{ij}$ is the actual fitted distance between the ith and jth points in the ordination, and $\text{f}(d_{ij})$ is a monotonically increasing function of the original distances (Fig. 3.17). STRESS represents the extent to which the rank

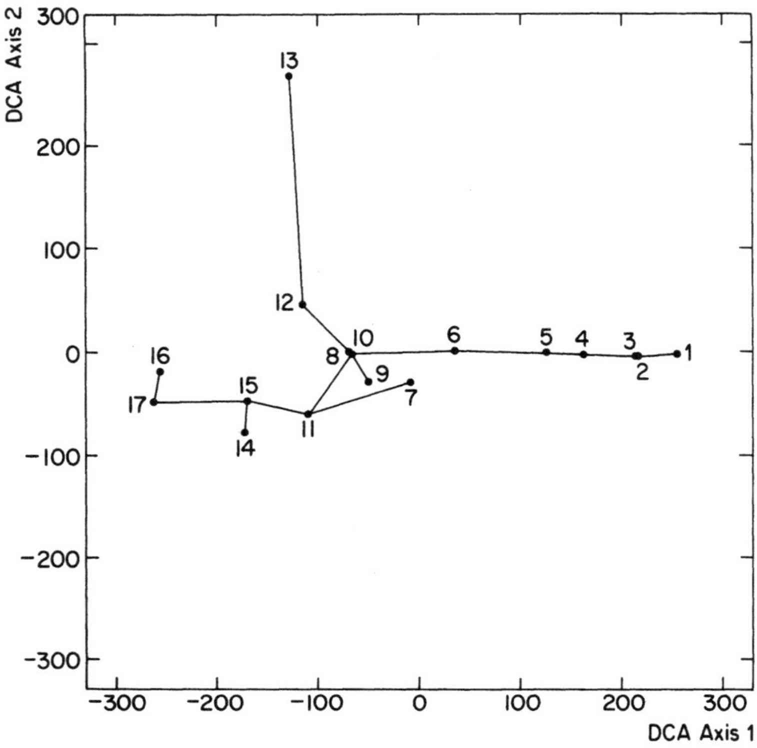

Fig. 3.16 **Ordination of the Steneryd meadow plots on the first two axes from a detrended correspondence analysis.**

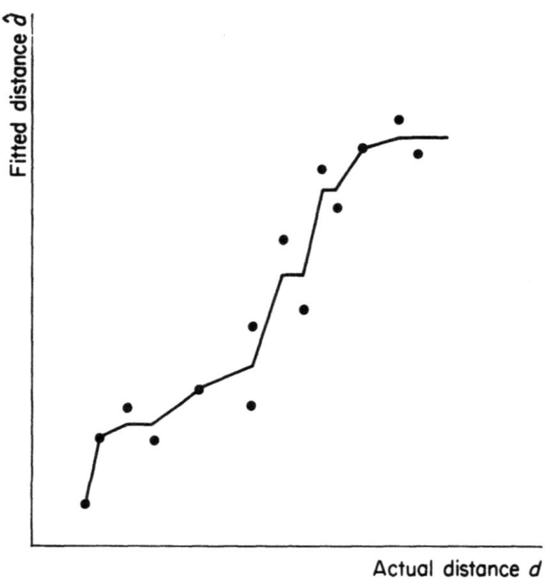

Fig. 3.17 Hypothetical plot of fitted distances among units, from an ordination, against their actual distances, showing the fitted least squares monotone regression line.

order of the values $\hat{d}_{ij}$ disagrees with the rank order of the value d_{ij}, thus emphasizing that the non-metric solution depends wholly on the ranked order of the original distances. Note that with non-metric methods it is customary for d_{ij} to refer to the actual distance between the ith and jth units, rather than their squared distance, as we have used elsewhere.

To minimize STRESS an iterative procedure is adopted: an initial solution is found and its STRESS calculated; the points are moved to reduce the STRESS and this procedure is repeated until no further reduction can be found; this is taken to be the best fitting ordination. A non-metric ordination is completely successful in representing the data, i.e. has STRESS zero, if the original distances are in the same ranked order as the distances among the points on the ordination; the monotonic function then passes through every point on the graph of $\hat{d}_{ij}$ against d_{ij}.

Several computer programs are available for non-metric ordination but care is needed in interpreting results since different programs use different procedures and may sometimes produce a final solution which is only a local minimum. The program KYST used for the examples in this book takes as its initial solution the principal coordinates ordination, and also performs a principal components analysis on the final solution to express this relative to its principal axes.

As already mentioned, non-metric ordination may have advantages for data where the scale of measurement has little intrinsic meaning or where some distances are unusually discrepant; it can also easily handle missing values by omitting these from the summation of equation (3.28). However, the methods also have several disadvantages. One of these is that the procedures are iterative and cannot guarantee to reach the global optimum. A second disadvantage concerns the relationship between solutions of different dimensionalities. With principal coordinates analysis the r ($<k$) dimensional solution is simply the first r dimensions of the solution in k dimensions. This is not the case with non-metric ordinations; here each solution must be computed afresh and will not generally relate to solutions of other dimensionalities. In choosing an appropriate dimensionality for an ordination, it may be helpful to plot STRESS for different numbers of dimensions; a sudden flattening of this curve indicates that there is no advantage in moving to the higher-dimensional configuration. A further problem of non-metric methods is that there are two different ways of treating tied (equal) distances: either they can be represented by equal rank values, or they can be considered as different and placed in the best order to minimize STRESS. This is unlikely to cause a problem with quantitative abundance data, but it may do so if distances are calculated from a small number of binary values (e.g. when sites are classified by the presence or absence of a small number of species) since then there may only be a small number of distinct distance values that can possibly occur.

Figure 3.18 shows the two-dimensional non-metric ordination of the Park

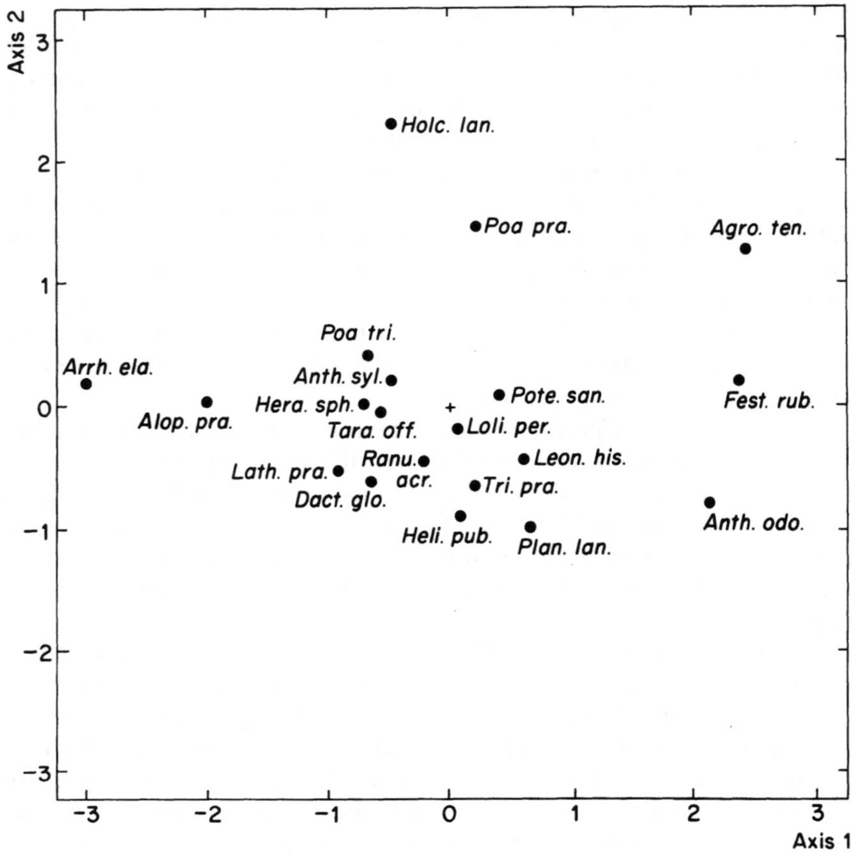

Fig. 3.18 Non-metric ordination of the Park Grass species. Minor species, not shown, are concentrated about the origin.

Grass flora based on the Euclidean distance matrix used for the principal components ordination of Fig. 3.5. The two ordinations are seen to be very similar, a fact explained by the close linear relationship shown in Fig. 3.19 between the actual distances among the species and the fitted distances from Fig. 3.18. The two points lying apart from the general pattern are found to represent the distances between the grass species *Festuca rubra* and the species *Agrostis* and *Anthoxanthum*. The relative positions of these species are poorly represented in the two-dimensional plot and, when the analysis was repeated for three dimensions, *Festuca* was found to be widely separated from all other species along the third axis. The fitted distances for the three-dimensional configuration more closely mirrored the actual distances and STRESS was reduced from 0.107 to 0.064.

In our view, non-metric ordination methods rarely have advantages over

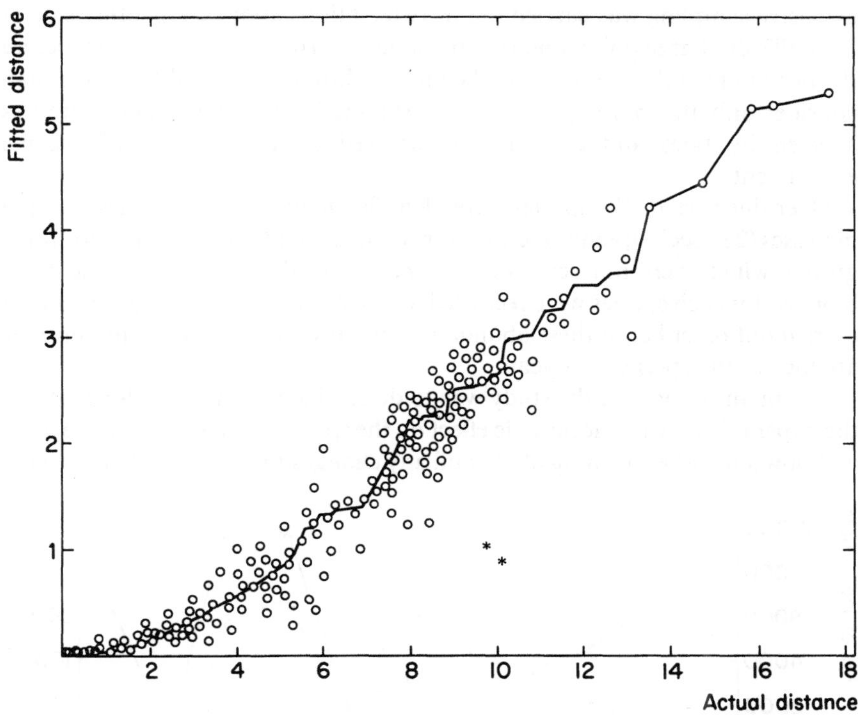

Fig. 3.19 Plot of actual distances among the Park Grass species against their fitted distances from the two-dimensional ordination of Fig. 3.18. Fitted line is the least squares monotone regression. The two starred points are referred to in the text.

metric methods and lead to greater problems of computation and interpretation: similar conclusions were reached by Gauch, Whittaker and Singer (1981). Consequently we are unable to recommend the general adoption of non-metric methods to ecologists.

3.8 CASE STUDIES

We conclude this chapter with two examples which illustrate further the application of ordination methods in different biological areas.

3.8.1 Monitoring pollution effects on the benthic fauna of a Scottish loch

Pearson (1975) describes the results of a detailed survey of changes in the benthic fauna of a sea-loch system on the west coast of Scotland following the introduction of effluent from a pulp and paper mill into the system. The most extensive sampling was carried out at two stations about 2 km either side of the mill's discharge point, and covered a period from 1963, three years before the beginning of the operation of the mill in March 1966, until 1973. We shall consider only the more inland station (Station 2) where the waters were less affected by tides and consequently showed a greater effect of organic enrichment.

Over this period, 128 species were identified mainly from three groups: the molluscs (22 species); echinoderms (9 species); annelids (94 species); plus three groups which were not separated to species. Both the numbers and total biomass for each species were recorded for each year, so any analysis could be carried out on either of these abundance measures, or simply the presence or absence of the species in a year.

The main purpose of the study was to determine whether the effluent from the paper mill had any noticeable effect on the species composition in the loch, and how soon after opening of the mill any changes took place. Changes were

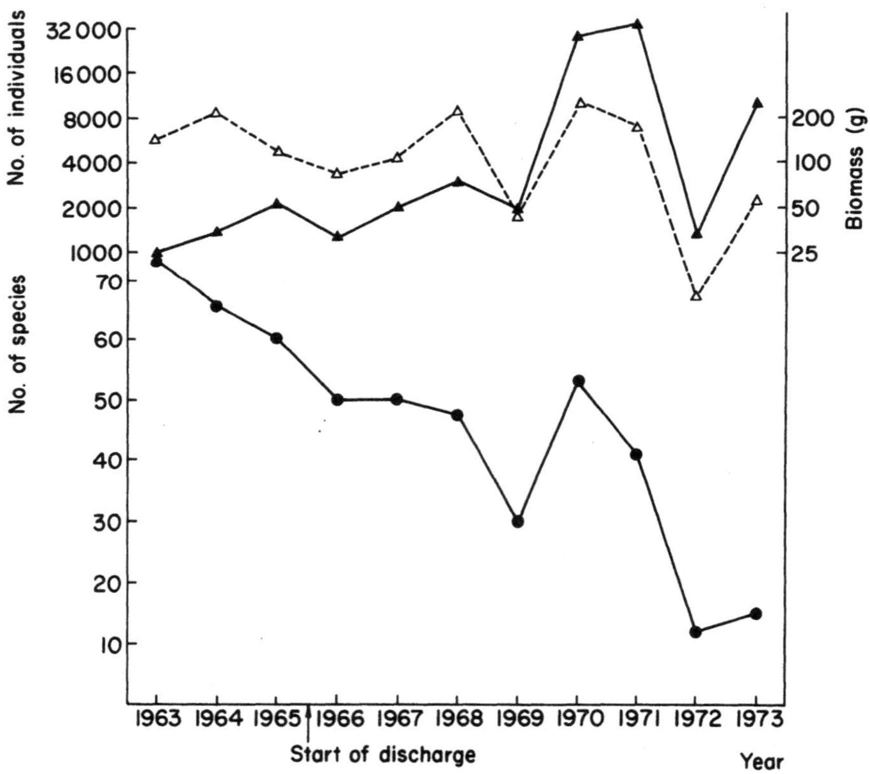

Fig. 3.20 Changes in the number of species (solid circles), number of individuals (solid triangles) and total biomass (open triangles) of samples from Loch Eil, Scotland, from 1963 to 1973. Effluent discharge from a paper and pulp mill close to the sampling station began in 1966 and resulted in a reduction in oxygen content of the upper waters of the loch (Pearson, 1975).

observed in the numbers of species, numbers of individuals, and total biomass of the samples (Fig. 3.20), but these were somewhat erratic and could not easily be attributed to pollution. A correspondence analysis carried out on the log numbers of each species in each year has a dominant first root; the ordination along this axis suggests that major changes in fauna did not occur until 1968, two years after the opening of the mill, but were still taking place at the end of the study in 1973 (Fig. 3.21(a)). A very similar ordination of samples is

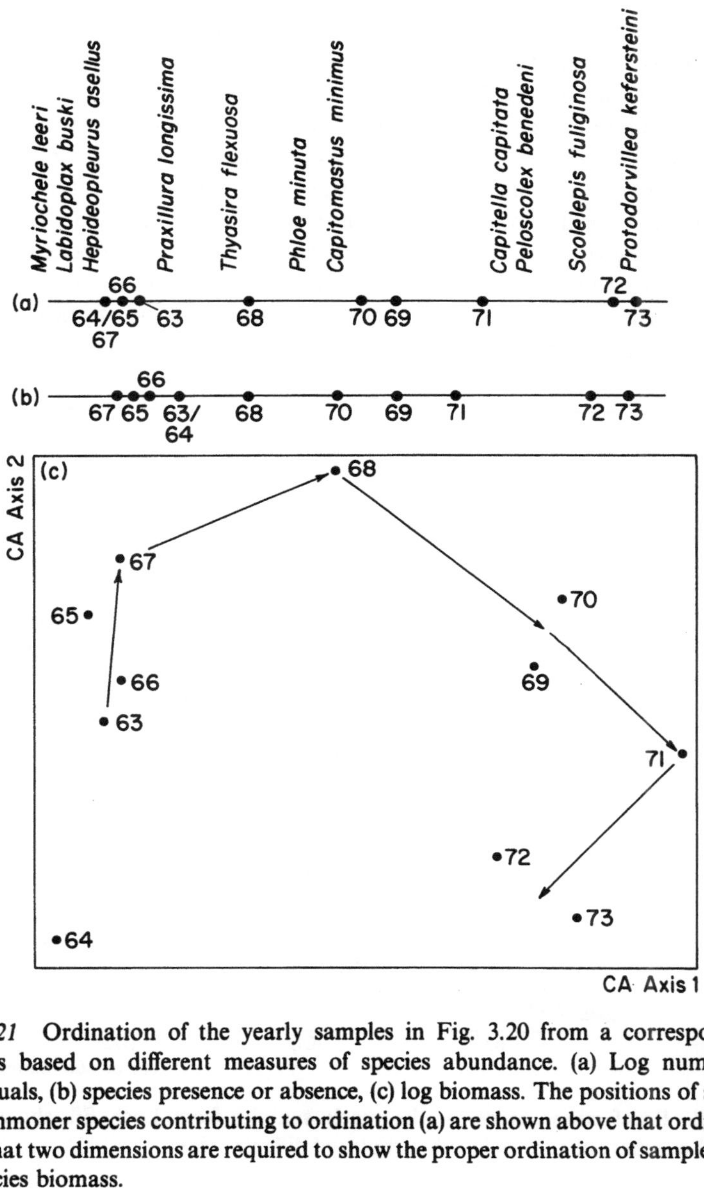

Fig. 3.21 Ordination of the yearly samples in Fig. 3.20 from a correspondence analysis based on different measures of species abundance. (a) Log numbers of individuals, (b) species presence or absence, (c) log biomass. The positions of some of the commoner species contributing to ordination (a) are shown above that ordination. Note that two dimensions are required to show the proper ordination of samples based on species biomass.

obtained when species abundance is denoted only by its presence or absence (Fig. 3.21(b)), but two dimensions are required to obtain the equivalent ordination from a correspondence analysis based on total species biomass (Fig. 3.21(c)).

Simple counts of species or individuals (Fig. 3.20) were shown to be ineffective as indicators of pollution when taken individually: the steady decline in species numbers is disturbed in 1970 and 1971, and again in 1973, when there was a large increase in overall population density. If species number is to be used as a diversity indicator, the samples must first be adjusted to have the same number of individuals. Following the method of Smith and Grassle (1977), we have calculated the expected numbers of species in sub-samples of different sizes for each year and produced the rarefaction curves (Sanders, 1968) of Fig. 3.22. The ordering of the yearly samples is remarkably consistent for sub-sample sizes within the range 100–1000 individuals, and there appears to be a steady reduction in species diversity in the loch from the beginning of the survey, including the three years prior to the operation of the mill. It is interesting that these rarefaction curves, which ignore species names, produce a clearer separation and a more consistent ordering of yearly samples than the correspondence analysis ordination. However, without an unpol-

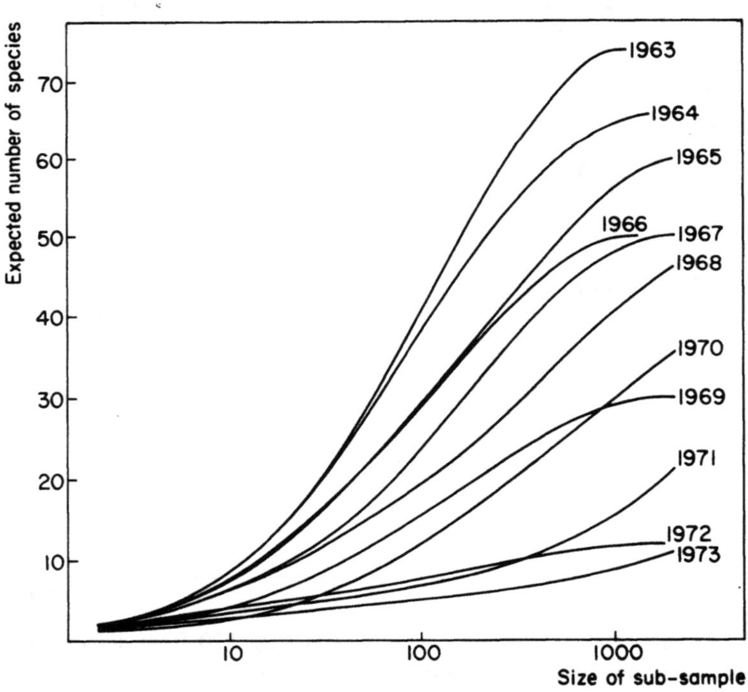

Fig. 3.22 Rarefaction curves for the yearly samples from Loch Eil, showing the expected numbers of species in sub-samples of different sizes.

luted control population for comparison, it is impossible to separate the effects of the mill's discharges from long-term natural changes in the population.

3.8.2 Spatial and temporal variation of aphid species in UK

In this example we consider the use of biplots for studying species-by-environment interactions. The approach was used by Kempton (1984), largely in the context of crop varieties. Here we apply the method to the abundances of 46 aphid species caught at 14 sites in the UK over a seven-year period from 1971 to 1977 (Taylor *et al.*, 1981). To investigate the sources of interactions of the species with the sites and years, we apply a biplot analysis to the full (46 × 98) species-by-environments table of aphid log abundances, doubly centred about their species and environment means. Figure 3.23 gives the principal components biplot with species points and environmental vectors separated for clarity. On the environmental plot (Fig. 3.23(b)), only the individual year points for Rothamsted are shown: these give rise to an average site vector (the mean of the seven pairs of coordinates) as shown, and the average vectors for the other sites and for years are obtained similarly. The first axis primarily separates northern from southern sites, while the second axis separates the hot dry summers of 1975 and 1976 from other years.

Axis 1 of the species plot (Fig. 3.23(a)) therefore identifies *Phorodon humuli* (hop aphid) and *Brevicoryne brassicae* (cabbage aphid) as predominantly southern species, while the two grass aphids, *Rhopalosiphum insertum* and *Rhopalosiphum padi*, occur with greater relative abundance in the north of the UK. The successive years 1975 and 1976 had respectively the lowest and highest overall total catches of aphids for the seven-year period but, from axis 2, several major pest species, including *Brevicoryne brassicae*, *Sitobion avenae* and *Myzus persicae*, had relatively high abundances over the whole of the UK in both years, while *Aphis fabae* and *Rhopalosiphum insertum* were present in relatively smaller numbers than other years. The third principal axis again separates average site vectors, discriminating broadly between western and eastern sites. Hence a biplot of axis 1 and axis 3 (Fig. 3.24) positions the sites roughly according to their geographical location. The anomalous position of the Hereford site to the 'south' of Wye, rather than in its true geographical position north of Long Ashton, arises from its large catches of the hop aphid, *Phorodon humuli*: Hereford and Wye lie within the two major hop-growing areas of the UK. The biplot predicts that this aphid has high abundance at both sites, but low abundance at sites in the north and far west.

In Fig. 3.24, the year vectors, although not shown, are clustered close to the origin: this biplot thus emphasizes differences in species' spatial distribution which are constant from year to year. Similarly, axis 2 describes yearly differences in the species' distributions and make little distinction among sites. Table 3.12 shows how the overall species × environment interaction sum of squares and its separate components due to species × sites, species × years and

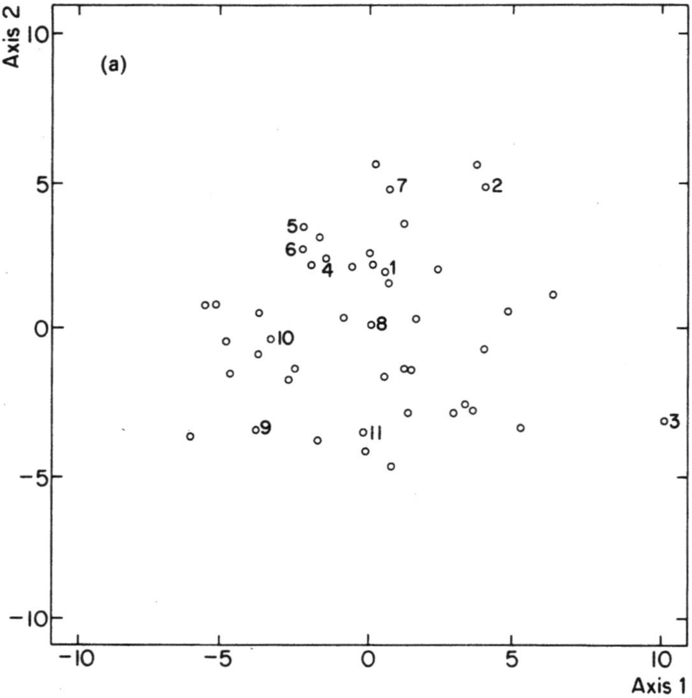
(a)
Axis 2
Axis 1
10
5
0
−5
−10
1
2
3
4
5
6
7
8
9
10
11

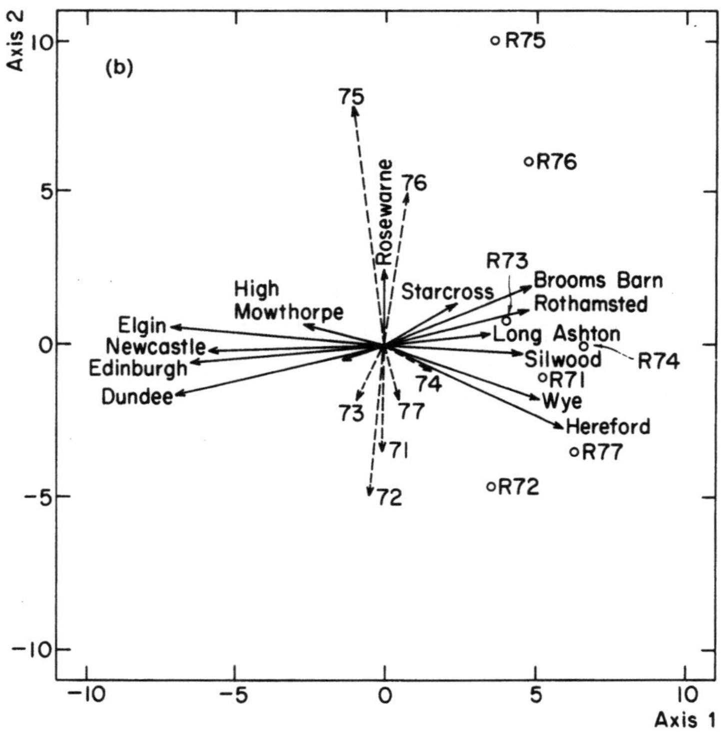
(b)
Axis 2
Axis 1
R75
R76
R73
R74
R71
R77
R72
75
76
73
77
74
71
72
Rosewarne
High
Mowthorpe
Starcross
Brooms Barn
Rothamsted
Long Ashton
Silwood
Wye
Hereford
Elgin
Newcastle
Edinburgh
Dundee

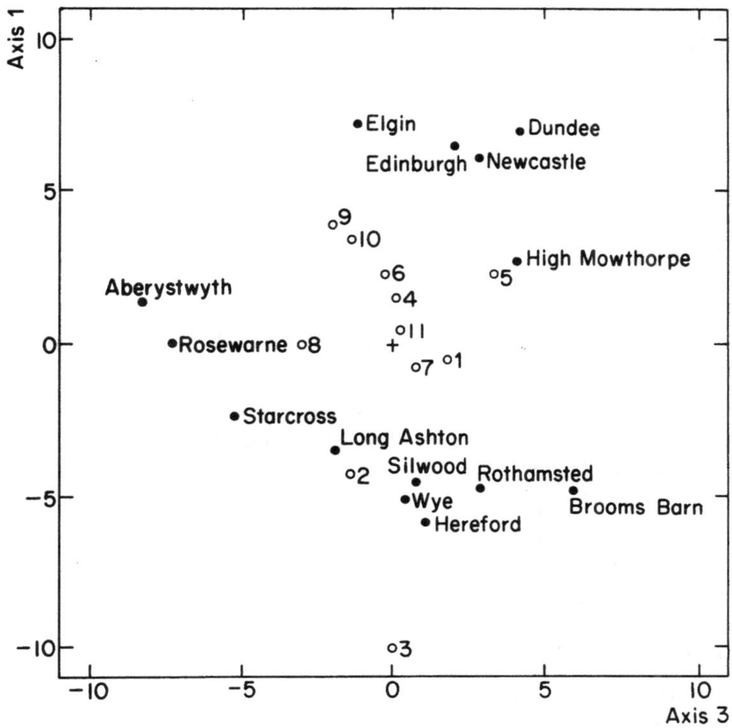

Fig. 3.24 Biplot for axes 1 and 3 showing the important aphid species (coded as in Fig. 3.23) and the mean positions of the 14 sites. (Note that axis 1 is here drawn vertically.)

Fig. 3.23 Biplot for 46 aphid species sampled in 98 environments (at 14 sites over 7 years).

(a) PCP plot for species: those of particular agricultural importance are coded as follows:

1. *Acyrthosiphon pisum*,	2. *Brevicoryne brassicae*,
3. *Phorodon humuli*,	4. *Myzus persicae*,
5. *Metopolophium dirhodum*,	6. *Macrosiphum euphorbiae*,
7. *Sitobion avenae*,	8. *Sitobion fragariae*,
9. *Rhopalosiphum insertum*,	10. *Rhopalosiphum padi*,
11. *Aphis fabae* grp.	

(b) Plot of environmental vectors. Solid and dashed lines represent the mean site and mean year vectors respectively. Points for individual years for the Rothamsted site are coded R71–R77.

Table 3.12 Proportion of each component of the species × environment interaction in log-abundances of aphid species caught at 14 sites over 7 years which is described by the four principal axes of a PCP.

	% of total species × environment interaction sum of squares					
Interaction component	*Total*	*Axis 1*	*Axis 2*	*Axis 3*	*Axis 4*	*Axes 1–4*
Species × site	43.8	21.1	1.2	5.7	1.6	29.6
Species × year	25.5	0.7	10.8	1.1	4.1	16.7
Species × site × year	30.7	0.9	3.1	1.4	2.0	7.4
Total	100.0	22.7	15.0	8.3	7.8	53.8

species × sites × years, is partitioned among the first four PCP axes. While around two-thirds of both the species × sites and species × years interaction sum of squares is described by the four principal axes, they account for less than a quarter of the species × sites × years interaction.

Principal components analysis applied to the full species by environments table has identified the major components of species × sites and species × years interactions but not the more complex three-factor interaction of species with individual sites in individual years. In some situations the three-factor interaction will have a large noise component and thus be of little interest. The individual two-factor interactions may then be summarized and displayed directly by analysing the separate two-way tables of species-by-sites and species-by-years means, $Y_{ij.}$ and $Y_{i.k}$, where Y_{ijk} is the log abundance of species i at site j in year k.

The close relationships between the ordination of sites in Fig. 3.24 and their geographical location suggests that abundance of the aphid species about their means could be described by a bivariate regression on grid reference:

$$Y_{ij.} - Y_{i..} = a_{i1}(W_{j1} - W_{.1}) + a_{i2}(W_{j2} - W_{.2}) + e_{ij}, \qquad (3.29)$$

where W_{j1}, W_{j2} are the northerly and easterly grid reference for the jth site and a_{i1}, a_{i2} are the corresponding regression coefficients for the ith species. Kempton (1984) found that this model described 51% of the variation in species abundances across sites. The two multiplicative terms in equation (3.29) each involve the product of a term for species and sites, and hence the expected abundances $E(Y_{ij.} - Y_{i..})$ may themselves be displayed as a biplot. Thus, if we add the points representing the site grid references $(W_{j1}, W_{j2}), j = 1, \ldots, p$, to the plot of species regression coefficients $(a_{i1}, a_{i2}), i = 1, \ldots, n$, the inner product of W_j and a_i gives the predicted abundance of species i at site j. The geographical locations may now be extended to include a map of the UK, so allowing the relative abundance of each species to be predicted in any region (Fig. 3.25). (Remember that when interpreting this biplot it is the size of the inner product that is important, not the appearance of a species point within a particular region in the figure; this is affected by the relative scales chosen for plotting species and locations.) The positions of species and sites in Fig. 3.25

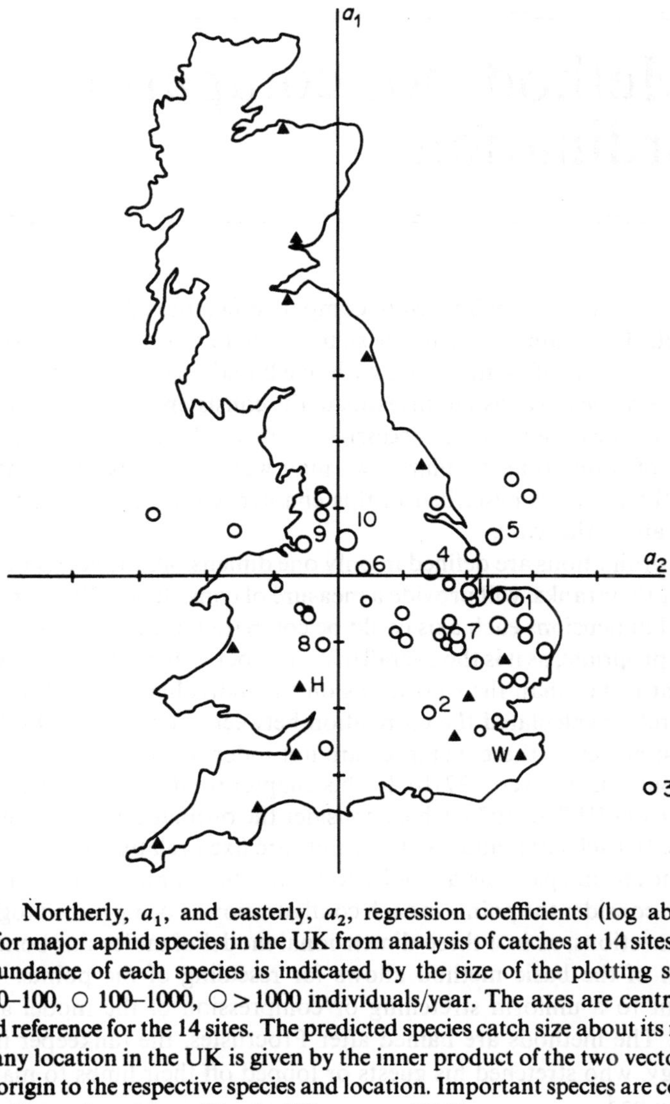

Fig. 3.25 Northerly, a_1, and easterly, a_2, regression coefficients (log abundance/100 km) for major aphid species in the UK from analysis of catches at 14 sites (▲). The mean abundance of each species is indicated by the size of the plotting symbol; ○ 1–10, ○ 10–100, ○ 100–1000, ○ > 1000 individuals/year. The axes are centred on the mean grid reference for the 14 sites. The predicted species catch size about its mean (log scale) at any location in the UK is given by the inner product of the two vectors drawn from the origin to the respective species and location. Important species are coded as in Fig. 3.23. (Reproduced from Fig. 6 of Kempton, 1984.)

are broadly in accord with the principal components biplot of Fig. 3.24. However, the very low abundances of the hop aphid at extreme westerly sites produces a large positive regression on easterly grid reference. Consequently the abundance at Hereford, which now occupies its correct geographical location on the plot, is very much underestimated by the inner product formula. Despite this, both biplot methods, working through regression or principal components, have been largely successful in describing succinctly the pattern of variation of aphid species over sites and seasons.

4 Methods for comparing ordinations

In some situations it is of interest to compare different ordinations of the same set of units. For example, we may wish to compare an ordination of sites based on the abundance of plant species with one based on environmental features; alternatively, we may wish to investigate the consistency of ordinations of the same set of sites obtained in successive years, or the consistency of different methods of ordination. Similarly, we may want to compare ordinations of species: the methods described in this chapter would, of course, be equally appropriate in this case.

If two ordinations are defined in only one dimension, the correlation of site scores, or their ranks, will provide a measure of consistency. With ordinations in several dimensions each axis could be compared in turn in this way: this is rarely appropriate, as it is the overall distances between points (sites) which are important rather than their positions on the individual axes. Many authors have therefore calculated the correlation between the two sets of inter-point distances; however, these distances are not independent so their correlation can be spurious (Gower, 1971b). In this chapter we describe several methods due to Gower (1971b, 1975) which consider the ordinations as configurations of points. In each configuration the points are fixed relative to each other but can be moved in space, as a whole, to fit another configuration – it may be helpful to consider the points in each configuration as being joined together by rigid connections, as in a three-dimensional model of molecular bonding. An extension of the basic method allows for rescaling of the points, which is equivalent to a uniform stretching or compression of the model about its centroid. The methods are named after Procrustes, the innkeeper in Greek mythology, who stretched his guests or lopped off their limbs to make them match the bed.

4.1 PROCRUSTES ROTATION

Procrustes rotation deals with two configurations of points (ordinations) representing the same set of n units (sites). One configuration is taken as fixed; the other is moved to match it as closely as possible. Initially, two types of movement which preserve distances among the points are allowed: translation, i.e. shifting the origin of the coordinate axis; and rotation of the axes. These operations are shown in Fig. 4.1(a), (b) and (c) and are chosen to

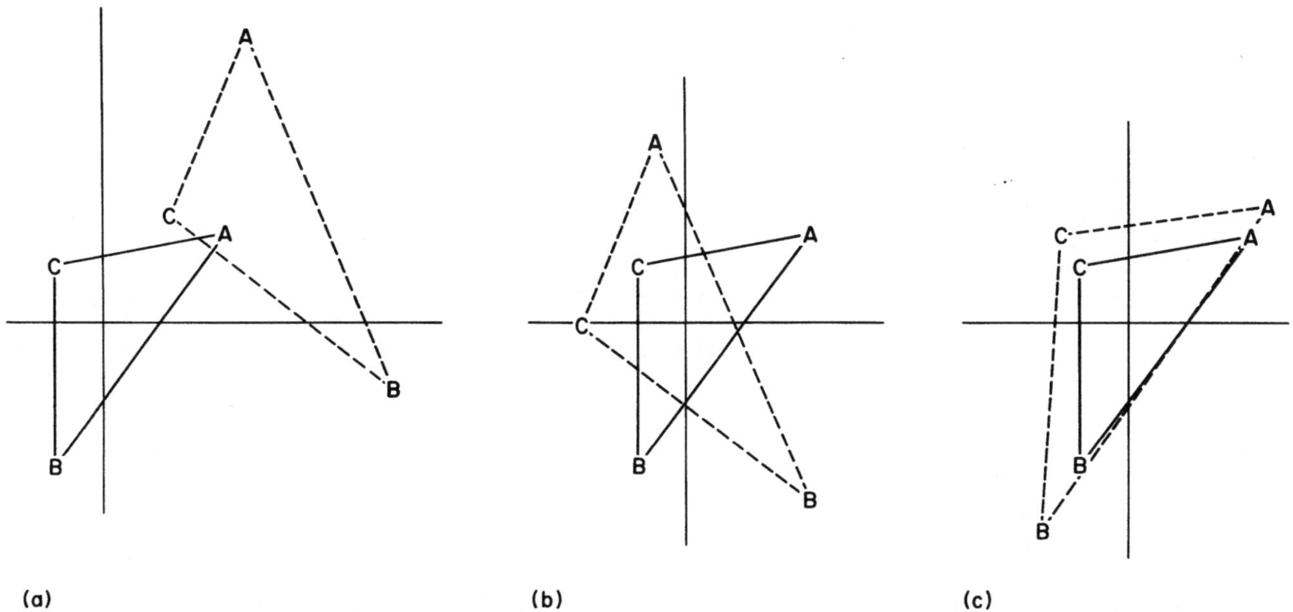

Fig. 4.1 Example of Procrustes rotation analysis: (a) original configurations of three points; (b) configurations after joint centring; (c) configurations after centring and rotation.

minimize the sum of squared distances of the transformed points from the respective points of the fixed configuration.

Algebraically, the two configurations of points in r-dimensional space are given by the $(n \times r)$ matrices $\mathbf{X}$ and $\mathbf{Y}$. If the two ordinations differ in their number of dimensions, then the smaller can simply be padded out with columns of zeros to match the size of the larger. Taking $\mathbf{Y}$ to be the fixed configuration, the coordinates of $\mathbf{X}$ are translated and rotated to a set of new coordinates $\mathbf{Z}$ such that the sum of squared distances

$$m_{XY}^2 = \text{trace}((\mathbf{Y}-\mathbf{Z})'(\mathbf{Y}-\mathbf{Z}))$$

is a minimum. It can be shown that the best translation is to make the centroid of $\mathbf{Z}$ coincide with that of $\mathbf{Y}$. This is conveniently achieved by a preliminary centring of $\mathbf{X}$ and $\mathbf{Y}$ so that both sets of points have their centroid at the origin. Now the required rotation of $\mathbf{X}$ to $\mathbf{Z}=\mathbf{XH}$ is given by the $(r \times r)$ orthogonal matrix

$$\mathbf{H}=\mathbf{VU}',$$

where $\mathbf{U}$ and $\mathbf{V}$ are given by the singular value decomposition of

$$\mathbf{Y}'\mathbf{X}=\mathbf{USV}'.$$

It sometimes occurs that two ordinations appear similar but one is on a larger scale than the other. For example, in Fig. 4.1 the configuration of points given by $\mathbf{X}$ is seen to be somewhat larger than that given by $\mathbf{Y}$. In this case a parameter p may be included in defining the transformation $\mathbf{Z}=p\mathbf{XH}$, so that the scales of $\mathbf{Z}$ and $\mathbf{Y}$ are commensurate; this is particularly appropriate when the scales are arbitrary, as is the case for many ordination methods. Scaling causes no immediate difficulty: the optimal translation and rotation are defined as before, and the value of the scaling parameter is then

$$p=\text{tr}(\mathbf{XHY}')/\text{tr}(\mathbf{XX}').$$

Often the choice of which configuration is to remain fixed is arbitrary and we then need to consider what is the result of rotating $\mathbf{Y}$ to match $\mathbf{X}$. If there is no rescaling the desired rotation is given by the inverse orthogonal matrix $\mathbf{H}'$ and the goodness-of-fit statistic is unchanged, $m_{XY}^2 = m_{YX}^2$. However, if the scaling parameter p is included, $m_{XY}^2 \neq m_{YX}^2$. This latter difficulty may be overcome by a preliminary rescaling of both matrices, after initial translation, to have unit sum of squares: $\text{tr}(\mathbf{Y}'\mathbf{Y})=\text{tr}(\mathbf{X}'\mathbf{X})=1$. Alternatively, both configurations may be rotated to fit a common configuration: this method may be extended for joint comparison of several configurations and is described in Section 4.2 under its title of 'generalized Procrustes analysis'.

4.1.1 Example of Procrustes rotation

In Section 3.2.2 we interpreted the two-dimensional ordination of Park Grass plots, obtained from a principal components analysis (PCP) based on species log abundances (Fig. 3.6), as showing two environmental gradients of soil pH and plot productivity. We now consider how closely the plot scores from PCP are described by these two environmental variables. Figure 4.2(a) shows the configuration of plots based on their soil pH and yield of dry matter, with both axes standardized to have zero means and unit standard deviations. Figure 4.2(b) shows this configuration rotated and scaled to fit the two-dimensional PCP. The rotation matrix (Table 4.1) describes an initial reflection of the first axis (pH) about its mean, as indicated by the minus signs in the first row, followed by an anticlockwise rotation of both axes through an angle of 66°. Table 4.1 also gives the overall fit of the yield/pH configuration to the PCP scores. The total sum of squares of PCP scores about the origin in the two-dimensional ordination is 424 (the sum of the first two eigenvalues), and about 65% of this is described by pH and yield.

Table 4.1 Results of a Procrustes rotation of the yield/pH scatter diagram on to the two-dimensional PCP for Park Grass plots (Fig. 4.2).

$$\text{Rotation matrix } \mathbf{H} = \begin{bmatrix} -0.411 & -0.912 \\ -0.912 & 0.411 \end{bmatrix}$$

Scaling parameter $p = 1.97$

	Sum of squares
Fitted yield/pH configuration	286
Residual	138
Total for PCP ordination	424

The residuals for the individual field plots are shown in Fig. 4.2(b). The largest residuals are associated with two pairs of sub-plots 14a, d and 11/1a, b. 14a and 14d correspond to the limed and unlimed halves of the main plot with high sodium nitrate, $N_2^{\dagger}$ PKNaMg, and have high soil pH but only moderate yield. In contrast, plots with similar species composition, in particular the limed ammonium nitrate plots 9a and 9b, which appear close to 14a and 14d on the PCP plots, have both high pH and high yield, and smaller residuals. Similarly, plots 11/1a and b have similar species composition to 11/2a and b and their treatments differ only in the addition of sodium silicate. However, plots 11/1a and b have lower soil pH and this results in the larger residuals shown in Fig. 4.2(b).

Despite the presence of these aberrant values, the general fit of the pH/yield configuration to the PCP ordination is good and justifies identifying these two

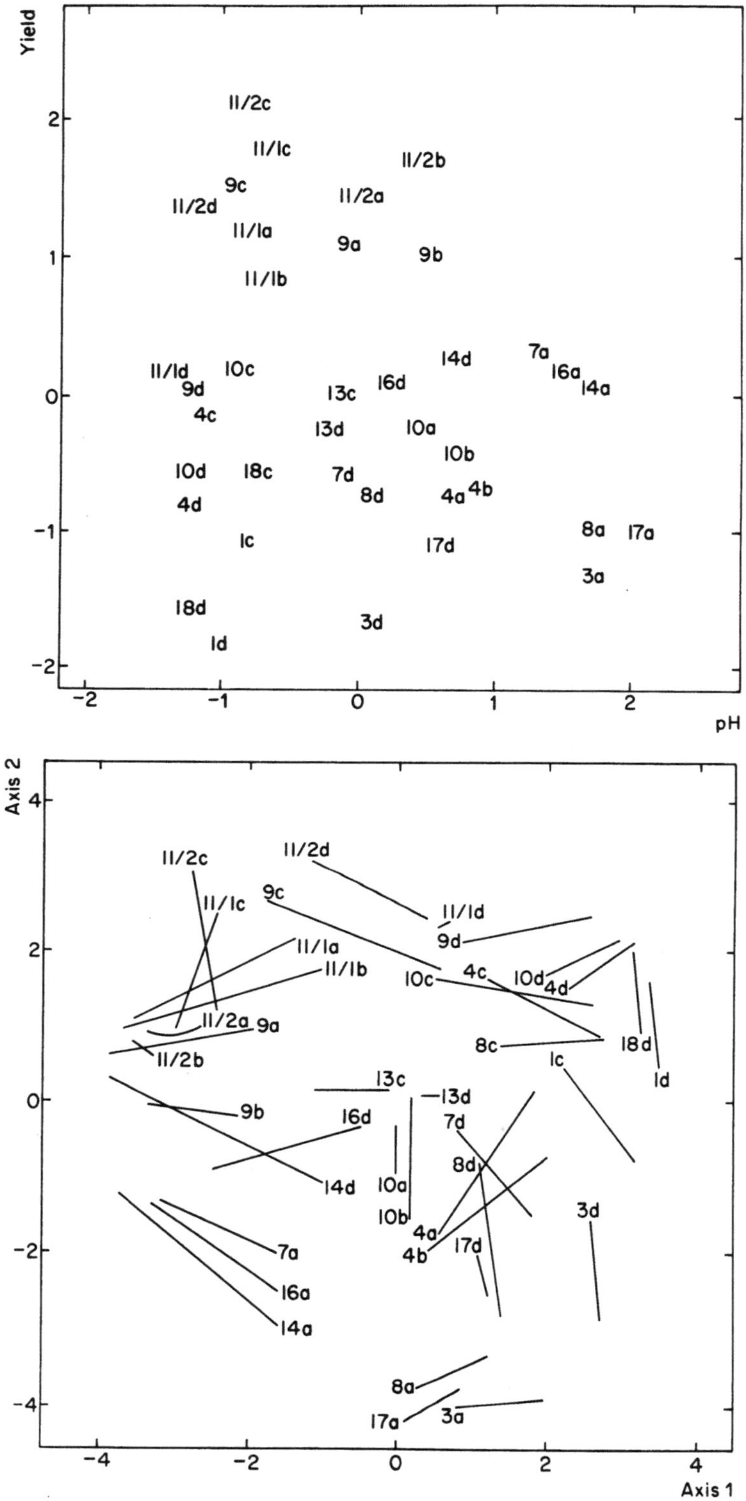

Fig. 4.2 Configuration of Park Grass plots for standardized yield and pH: (a) unrotated; (b) rotated, reflected and scaled to fit PCP ordination of Fig. 3.6. Lines connect labelled points for the rotated yield/pH ordination to their respective fixed points from the PCP.

environmental gradients as the major factors determining the species composition of the plots.

4.2 GENERALIZED PROCRUSTES ANALYSIS

This method, described by Gower (1975), can be used with any number of configurations. The basic idea is to find a consensus, or centroid, configuration so that the fit of the ordinary Procrustes rotations to this centroid, over all configurations, is optimal.

Again, we assume that each configuration has been translated to have its centroid at the origin; let us denote these initial configurations by $\mathbf{X}_i$ $(i=1, 2, \ldots, k)$, where there are k configurations. The final consensus configuration $\mathbf{Y}$ is the average of the result of rotating each $\mathbf{X}_i$ to $\mathbf{Y}$. Let $\mathbf{Y}_i$ denote these results, then

$$\mathbf{Y}_i = p_i\mathbf{X}_i\mathbf{H}_i,$$

$$\mathbf{Y} = \frac{1}{k}\sum_{i=1}^{k} \mathbf{Y}_i,$$

where p_i and $\mathbf{H}_i$ are the scaling and rotation parameters that minimize

$$m_i^2 = \mathrm{tr}[(\mathbf{Y}-\mathbf{Y}_i)'(\mathbf{Y}-\mathbf{Y}_i)].$$

Overall, we seek scaling and rotation parameters to minimize $m^2 = \Sigma m_i^2$. Clearly there is a trivial solution when all the scaling parameters are zero, giving zero values for all m_i^2. To avoid this, the method arranges that the overall sum of squares of the configurations is unchanged by the rotations, so that

$$\sum_i \mathrm{tr}(\mathbf{X}_i'\mathbf{X}_i) = \sum_i \mathrm{tr}(\mathbf{Y}_i'\mathbf{Y}_i) = \sum_i p_i^2\ \mathrm{tr}(\mathbf{X}_i'\mathbf{X}_i).$$

No simple solution to this problem exists and an iterative scheme must be used. Typically, we choose an initial centroid configuration and repeatedly update this from determination of rotations of each $\mathbf{X}_i$ to the current centroid (Gower, 1975; ten Berge, 1977).

The orientation of the final solution, given by the centroid $\mathbf{Y}$ and the configurations $\mathbf{Y}_i$ $(i=1,2,\ldots,k)$, is arbitrary. Having obtained a final solution it is convenient to rotate $\mathbf{Y}$ to its principal axes, using principal components analysis; this rotation is then also applied to the final individual configurations $\mathbf{Y}_i$.

4.2.1 Examining the consistency of ordinations from replicate samples

One of the reasons for using a multivariate ordination to identify major environmental gradients is that such ordinations should be more robust to sampling variations than those based on a single species. A similar rationale lies behind the preference for polythetic rather than monothetic methods of classification. The robustness of ordinations is seldom considered but may be investigated by taking a succession of replicate samples in time (Kempton, 1981) or space (Wilson, 1981) for each community.

Several problems arise when comparing the replicate ordinations. Firstly, the variation in species composition among the samples of any one replicate is due both to identifiable environmental variation and to unexplained, partly random, noise. In an ordination this noise component will mainly be assigned to lower dimensions, which consequently would not be expected to show any consistency from one replicate to another. Thus the experimenter must first decide on the dimensionality of the ordinations to be compared, and this will depend on an assessment of the number of environmental gradients which are displayed. While the same environmental gradients may be present in successive replicates, they may differ in their relative importance: thus the primary gradient in one year may, in a subsequent year, only be shown in the second or third dimension of an ordination. Hence the axes of the ordinations may need to be rotated before comparison. Axes may also need to be reflected about their origin since the sign of a gradient produced by an ordination is usually arbitrary. Finally, the scales of the ordinations may differ, either because of differences in the size of the samples (though this should cause no problems when using an ordination method based on standardized species abundances) or because of differences in the 'noise' level of the samples, and consequently in the proportion of the variation described by the first few dimensions of the ordination. These considerations make the generalized Procrustes method very suitable for comparing replicate ordinations.

In Section 3.4.2 we used canonical variate analysis to obtain an ordination of 14 sites in the UK, based on their yearly catches of 12 major moth species over the six-year period, 1969–74. Here we use generalized Procrustes analysis to find an overall ordination of sites and at the same time investigate the consistency of the ordinations obtained for individual years. The ordinations are from correspondence analyses which were based on the abundances of all 510 species caught over the six years. The two-dimensional ordinations (Fig. 4.3) show a closely similar pattern. Axis 1 separates northern sites (J, K, L, M, N) from southern sites (see inset to Fig. 3.12). Except for 1974, axis 2 gives a fairly consistent ordering of southern sites which may be related to the area of woodland within the trapping area: trap A is located within an area of undisturbed woodland while trap I is in an intensively cultivated area on an agricultural experimental station. However, the ordination for 1974 is radically different in axis 2 and identifies site L as being different from all the others. A generalized Procrustes analysis applied to the six two-dimensional ordinations emphasizes the difference of this particular ordination; the residual sum of squares m_i^2 for 1974 is larger than the total for all other years combined (Table 4.2(a)).

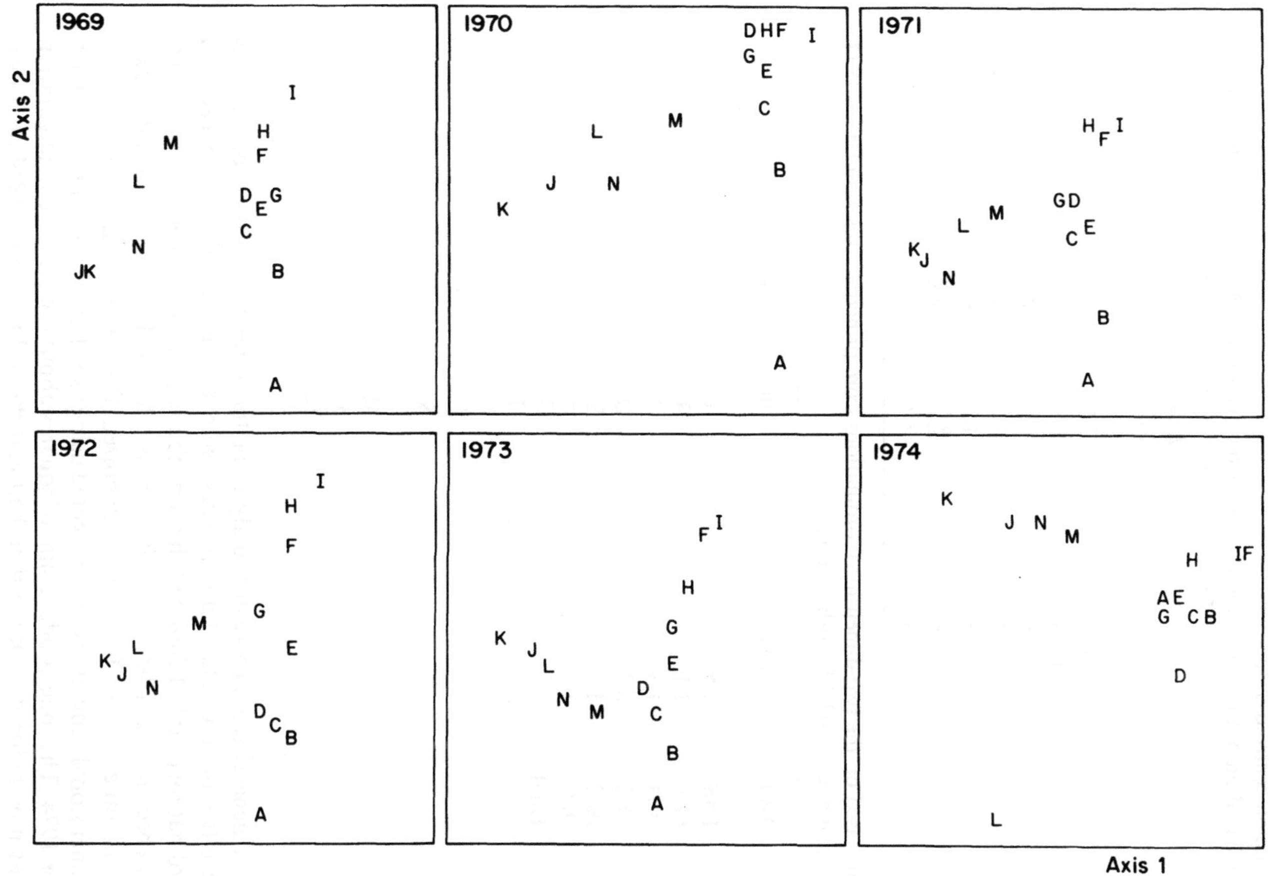

Fig. 4.3 Correspondence analysis results for 14 sites based on species abundance in each of six years (1969–74). For site codes see Table 3.6.

Table 4.2 Partition of the total residual sum of squares (expressed as a % of the overall sum of squares) for the individual yearly ordinations of moth-trap sites about a common centroid ordination.
(a) Based on ordinations from the first two or first three PCP axes for all years.

	Residual s.s.	
Year	*2-D*	*3-D*
1969	1.6	5.4
1970	1.9	3.4
1971	0.8	2.5
1972	1.6	3.1
1973	1.7	4.4
1974	8.0	3.3
Total	15.6	22.1

(b) Based on the two-dimensional ordinations from principal axes 1 and 3 for 1974, and axes 1 and 2 for other years.

Year	*Residual s.s.*
1969	1.5
1970	1.1
1971	1.3
1972	0.3
1973	1.7
1974	1.3
Total	7.2

Site	*Residual s.s.*
A	0.2
B	0.2
C	0.5
D	1.3
E	0.1
F	1.4
G	0.1
H	0.5
I	0.8
J	0.3
K	0.5
L	0.2
M	1.0
N	0.1
Total	7.2

A generalized Procrustes analysis of the three-dimensional ordinations resulted in a greater total residual sum of squares about the centroid ordination (Table 4.2(a)), but the contribution due to 1974 was reduced. This was because the ordination of sites on axis 3 for 1974 was similar to that on axis 2 for other years. Thus, the generalized Procrustes analysis of the two-dimensional ordinations was repeated using axes 1 and 3, rather than 1 and 2, for 1974. The total residual sum of squares about the centroid configuration was now reduced by more than half and the ordination for 1974 no longer appeared aberrant (Table 4.2(b)). Figure 4.4 shows this centroid configuration. The scores for the individual years have been added for three sites, A, D and K.

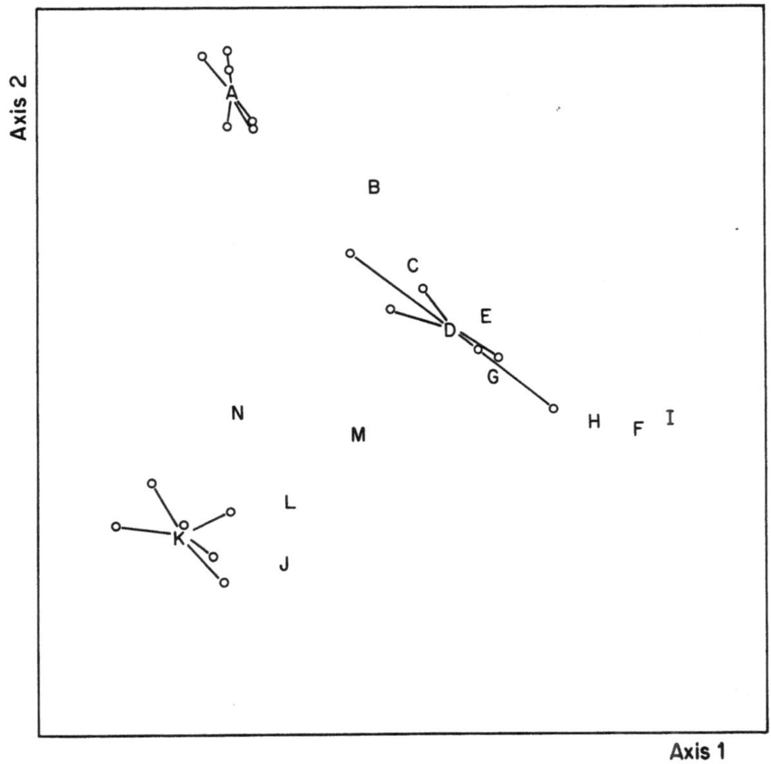

Fig. 4.4 Centroid configuration from a generalized Procrustes analysis of site ordinations in each of six years. Positions for individual years are shown joined to their centroid point for three of the sites.

Plotting the scores in this way can help to identify sites which show greater variation across years. This can also be shown by partitioning the total residual sums of squares into components for each site, as in Table 4.2(b).

4.3 COMPARING ORDINATION METHODS BY MULTIPLE PROCRUSTES ANALYSIS

We now use Procrustes analysis to compare the results of different ordination methods. Most previous comparisons (e.g. Gauch, Whittaker and Wentworth, 1977) have been based on simulated data which describe a single environmental gradient: a straightforward visual comparison of the results with those expected may then be quite satisfactory. However, in complex practical examples a more objective comparison is required; this is provided by Procrustes methods.

We shall compare the results of applying 12 different ordination methods to

Table 4.3 Ordination methods used and values of the m^2 statistics ($\times 1000$) from comparing pairs of ordinations.

Ordination methods

P1–P5, principal coordinates analysis using five different similarity measures:
- P1, simple matching coefficient for species presence/absence;
- P2, as P1, ignoring joint absences of species;
- P3, Manhattan metric on log relative abundances;
- P4, as P3, ignoring joint absences;
- P5, Euclidean metric on log relative abundances;

N1–N5, non-metric scaling (Section 3.7) based on the five similarity indices of P1–P5;
C1, correspondence analysis of species presence/absence;
C5, correspondence analysis of species abundance.

m^2 statistics

	P1	P2	P3	P4	P5	N1	N2	N3	N4	N5	C1	C5
P1	—											
P2	1	—										
P3	23	23	—									
P4	40	38	4	—								
P5	30	28	2	3	—							
N1	4	5	23	38	28	—						
N2	20	16	37	47	36	18	—					
N3	25	25	7	12	8	24	30	—				
N4	55	54	29	24	23	54	49	14	—			
N5	44	42	15	11	7	40	39	10	9	—		
C1	11	13	23	40	32	10	36	33	59	50	—	
C5	74	68	26	17	18	64	61	39	50	25	70	—

the 14 sites of the previous section based on their species catches over the whole six-year period. The methods are listed in Table 4.3. The 12 two-dimensional ordinations were each scaled to have unit sum of squares and were then compared in pairs by Procrustes analysis (Section 4.1). The values of the m^2 statistic (Table 4.3) indicate the degree of dissimilarity between the results of each pair of ordination methods, and certain patterns are immediately apparent.

The pairs of methods (P1, P2) and (P3, P4) give a very low value of m^2, indicating that the treatment of species' joint absences had little effect on the final ordination. This was surprising as the proportion of species comparisons classified as joint absences varied considerably across all pairs of sites (i.e. from 25% to 55%). The non-metric ordinations are mostly similar to their metric equivalents, but the metric ordinations P1 to P5, as a group, are more similar than the non-metric ordinations.

The m^2 statistics of Table 4.3 can be considered as squared distances. Hence the similarities between the methods may be represented in two dimensions by

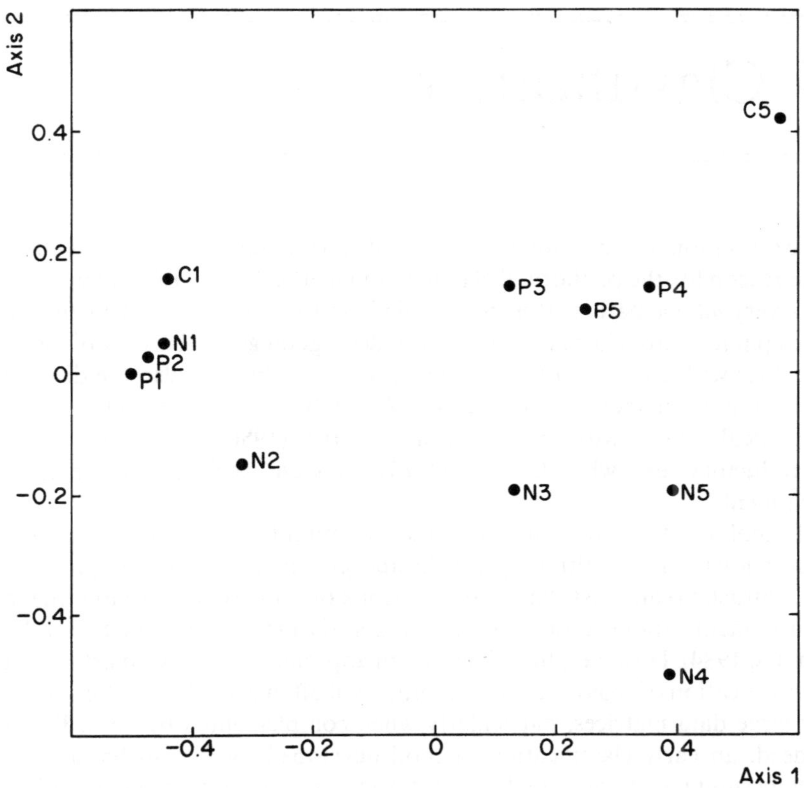

Fig. 4.5 Location of ordination methods on the two principal coordinates axes: these two axes represent 75% of the variation of the m^2 statistics in Table 4.3. For codes see Table 4.3.

a principal coordinates analysis (Section 3.5) of these values. The resulting display (Fig. 4.5) emphasizes the patterns of similarities noted above. Interestingly, the first axis separates methods based on quantitative abundance data from those based on species presence or absence, while the second axis separates metric from non-metric methods. The variation in the ordinations produced by different methods was far greater than the year-to-year variation experienced when using a single method (Section 4.2.1).

5 Classification

The formation of items into groups is a natural mental activity. This is possibly one reason for the plethora of classification methods available, many of them of a very *ad hoc* nature. It is impossible here to cover all the ideas that have been put forward: Cormack (1971) provides a good general review of the area, together with some useful cautionary comments (but see also the discussion following that paper by J. C. Gower). A number of books are also available that deal solely with classification: Everitt (1980) is a very readable introductory text, while Gordon (1981) gives an excellent, more extensive, treatment.

In ecology, it is now usual to consider community variation as being largely continuous, when partitioning into discrete groups may appear inappropriate. In contrast, taxonomists have tended to look on species as discrete entities and to use classifications to summarize their associations (Sneath and Sokal, 1973; Davies, 1984). However, this difference in approach is somewhat artificial and a classification of ecological communities will often provide a useful summary for large data matrices, particularly when complemented by an ordination. Indeed, an early classification method developed by Braun-Blanquet (and lucidly summarized by Gauch, 1982) starts by ordering the rows and columns of the species-by-sites matrix to bring together sample sites with a similar species composition, and species with a similar distribution over sites. As we saw in Chapter 3, this may be achieved using correspondence analysis or a similar ordination procedure (see Table 3.2).

Classification methods can themselves be classified as either hierarchical or non-hierarchical. Non-hierarchical methods merely assign each item to a group. With hierarchical methods the groups are themselves arranged into a hierarchy: thus any grouping of units into, say, k groups is part of a separate grouping into $(k-1)$ groups, and so on. There are two methods for forming a hierarchical classification: agglomerative and divisive. Agglomerative methods initially consider each unit as being a separate group and proceed by repeatedly combining the two closest groups until only a single group remains. Divisive methods start with all the units as one group and proceed by repeatedly dividing groups into two until all the units are separate.

In general, there are computational difficulties with many non-hierarchical and divisive hierarchical methods; this has led to agglomerative hierarchical methods becoming popular. However, agglomerative methods have been criticized by ecologists for concentrating initial attention on the similarities between individuals and small groups: in consequence, Gauch (1982) argues, the important major groupings of communities are not always robust. Nevertheless, these methods will often provide a useful and objective preliminary classification, particularly when associated with an ordination: we begin by describing them before considering the alternatives.

5.1 AGGLOMERATIVE HIERARCHICAL METHODS

Agglomerative hierarchical methods, often collected together under the title 'cluster analysis', operate on a matrix of similarities among a set of units. As shown in Chapter 1, similarities may be constructed from the more usual rectangular data matrix of sites by species using various possible measures of similarity or distance. The common basis of a similarity matrix makes it particularly attractive to combine the results of a cluster analysis with an ordination by principal coordinates analysis. This also has the benefit of highlighting aberrant clusters, as we demonstrate later.

Starting from the full inter-unit similarity matrix, all agglomerative methods begin by joining the two most similar units into a single group. The similarities of this group and all the other units are then calculated (this is where the methods differ). At the second stage the largest remaining similarity, among either two units or the new group and another unit, determines the next join. This process repeats until there are only two groups, or clusters, which are finally merged.

5.1.1 Single linkage

The most popular method of cluster analysis for general application is single linkage, also called nearest neighbour. The method is best explained by thinking of an ordination of the units in multidimensional space. The join that merges any unit with either another unit or a group involves the unit's nearest neighbour. In the first case the two close units join (in this situation the two units are mutual nearest neighbours); in the second the group includes the unit's nearest neighbour. The common term used for the method arises because it is the single links, between units and their nearest neighbours, that establish the clustering.

The numerical procedure for forming single linkage clusters may be described as follows:

(i) start with n 'groups', each containing just one unit;
(ii) join the most similar units, say the ith and jth, into a single group, so that there are now $(n-1)$ groups;
(iii) derive the similarity between this new group and every other unit k as the greater of the similarities between units i and k and units j and k;
(iv) join the two most similar 'groups', which will either be two individual units or one unit and the group formed in (ii);

(v) derive new similarities between the new group and all other groups, using the rule in (iii).

Then continue to combine the groups so that at each stage the number of groups is reduced by one and the similarity between two groups becomes the similarity between their closest members.

The procedure can be terminated at any stage. If continued to completion, the set of units will be linked together by their minimum spanning tree (Section 3.5.4). This is a set of links joining together units such that there are no closed loops and a path exists between all pairs of units. If the lengths of the links are equal to the appropriate inter-unit dissimilarities, the minimum spanning tree is such that the sum of the lengths of the links is a minimum. Alternative, more efficient, procedures exist for deriving the minimum spanning tree (Gower and Ross, 1969) and, once obtained, it may be used to emulate single-linkage clustering (in reverse) by progressively cutting links which exceed a certain dissimilarity.

Figure 5.1 shows the minimum spanning tree connecting light-trap sites plotted on the first two axes of the ordination derived in Section 3.4.2. Here the

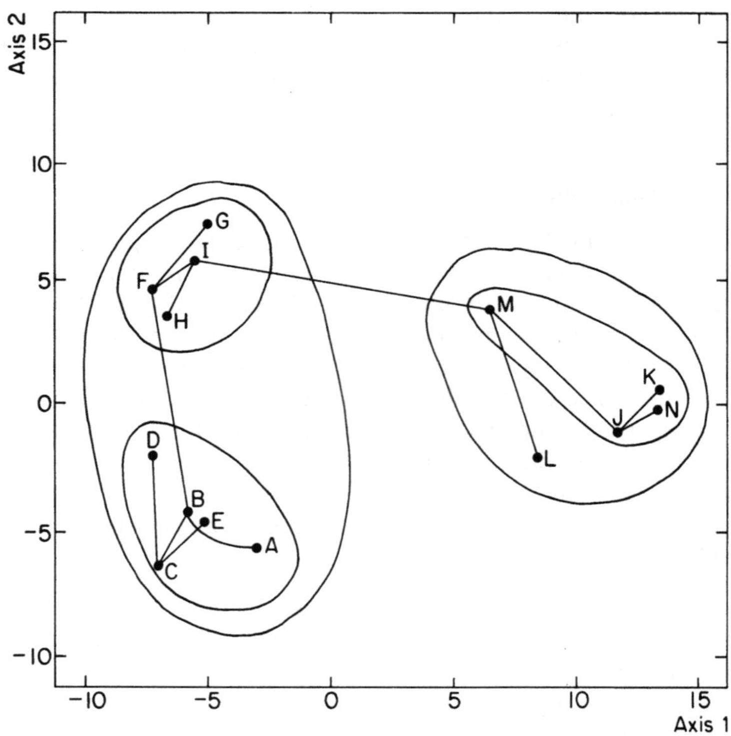

Fig. 5.1 Ordination of moth samples from 14 sites (as Fig. 3.12) connected by their minimum spanning tree and showing clustering at two hierarchical levels.

similarities among sites are calculated from the Mahalanobis squared distances. Cutting the largest link (I to M) of the minimum spanning tree produces two groups, representing sites from the north and south of the UK. At a higher similarity threshold, two more links are cut: B to F and L to M. The former of these separates predominantly woodland sites (A, B, C, D and E) from the other southern sites.

Single-linkage cluster analysis has the attractive property that computation is relatively straightforward. However, results from single-linkage analysis may be unsatisfactory because of the method's propensity for chaining: producing long, straggly clusters in the ordination space, in which many members have little in common but are linked together by a chain of intermediate units. (An example is given in Fig. 5.9.) When there is an overlap of clusters the method tends to produce a single cluster and a large number of unjoined units which are joined one at a time to the main group. Dissatisfaction with this behaviour led to other clustering methods being proposed.

5.1.2 Complete linkage

Complete linkage, or furthest neighbour, cluster analysis may be viewed as the opposite of single linkage. In this method, the similarity between unit k after units i and j have been merged is the smaller of the similarities between units i and k, and units j and k. This replaces operation (iii) for single-linkage analysis (see Fig. 5.2). The similarity between two groups is defined to be the similarity between their furthest members – hence the alternative name for the method. The requirement for all pairs of units within a group to exceed some minimum similarity means that the method tends to produce very compact spherical clusters; often one gets a large number of small clusters and a small number of slowly growing major groups.

5.1.3 Average linkage and centroid methods

Complete and single linkage can be considered as two extreme clustering criteria. We now describe two intermediate methods for calculating similarities between groups.

In the average linkage method, the similarity between unit k and the new group formed by the join of units i and j is the arithmetic mean of the similarities between units i and k and units j and k. In geometrical terms we can think of each cluster as having a single centre and the similarities between groups defined by the distances between centres: initially, there are n clusters with centres at the points identifying the individual units on the ordination; as the clustering proceeds the centres move so that, after any join, the new centre is half-way between the centres for the two merged groups. Using unweighted averages, as in this method, can lead to a large shift in the position of the

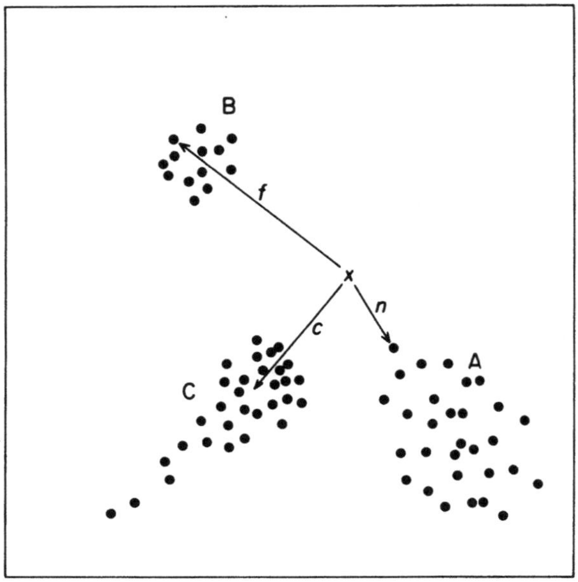

Fig. 5.2 Illustration of three alternative clustering methods for points described in two dimensions. x represents a unit to be assigned to one of the groups A, B or C. n, f and c are the shortest distances from x to the nearest neighbour, furthest neighbour and centroid of the three groups. Nearest-neighbour (single linkage) clustering assigns x to A, furthest neighbour (complete linkage) clustering assigns x to B, and centroid linkage clustering assigns x to C. (After Davies, 1984, Fig. 2.)

cluster centre when a single unit joins a large group. To avoid this some form of weighted similarity may be preferred, typically weighting by group size.

Centroid cluster analysis is again best thought of in geometrical terms. When combining groups, it differs from the average method in that the centre of the newly formed cluster is defined to lie at the centroid of the units in the cluster. This means that the inclusion of a single outlying unit is unlikely to affect the cluster centre as much as the unweighted average method. Both methods have advantages over single or complete linkage clustering in that they chain far less often but are not restricted to giving spherical clusters.

5.1.4 Further comparison of clustering methods

The final result from any clustering procedure will depend both on the initial choice of similarity measure used for comparing units (Section 1.4) and the choice of criterion for defining group similarity. It will also depend on the stopping rule for terminating the procedure (e.g. the number of groups desired).

As far as the choice for defining the group similarity measure is concerned, all the methods previously described in this section redefine the similarity $s_{k(ij)}$ between the kth unit (or group) and the newly merged ith and jth units (or groups) in terms only of the prior similarities, s_{ij}, s_{ik}, s_{jk}, between the three groups i, j and k. In terms of distances, or dissimilarities $d=1-s$, the new distance can be written

$$d_{k(i,j)}=a_i d_{ik}+a_j d_{jk}+b d_{ij}+c\left|d_{ik}-d_{jk}\right|, \tag{5.1}$$

where the parameters a_i, a_j, b and c depend on the clustering method used. The values of these parameters for the four methods mentioned above are shown in Table 5.1. The ability to redefine the group similarities in this way has computational advantages, as it means that the initial $(n\times n)$ similarity matrix need not be retained but can be overwritten as the clustering proceeds.

Table 5.1 Parameter values used for deriving group similarities (equation (5.1)) for different clustering methods.

Method	a_i	b	c
Single linkage	$\frac{1}{2}$	0	$-\frac{1}{2}$
Complete linkage	$\frac{1}{2}$	0	$\frac{1}{2}$
Average linkage	$\frac{1}{2}$	0	0
Centroid	$n_i/(n_i+n_j)$	$-n_i n_j/(n_i+n_j)^2$	0

n_i denotes the number of units in the ith group.

Varying a_i, a_j, b and c in equation (5.1) provides a family of clustering methods whose characteristics depend on the parameters. It is usual to include some constraints on the parameter values, for example $a_i+a_j=1$; Lance and Williams (1966) provide general guidelines.

It can readily be shown from equation (5.1) that the results of clustering by any of the methods described in this section will be unaffected by a linear transformation of the similarities: $s^*=us+v$. In addition, single and complete linkage results remain unchanged under any monotonic transformation of similarities. This means that results for any particular clustering method will be the same for a number of different similarity coefficients (e.g. for presence–absence data, both the Jaccard and Czekanowski coefficients should lead to the same clustering).

5.2 DIVISIVE HIERARCHICAL METHODS

In ecology, divisive methods of classification have been generally recommended in preference to agglomerative methods (Hill, Bunce and Shaw, 1975) as the classification of units (e.g. stands of vegetation) into broad categories is usually of far greater interest than the association among particular pairs. The simplest divisive methods operate only on binary data and are monothetic, i.e. each group division is based on the state of a single species. At each stage, the particular species used for division is chosen so that all units within each of the resulting subgroups are as similar as possible. Two criteria of similarity were proposed by Williams and Lambert (1959), who called the method 'association analysis', and Lance and Williams (1968). These authors also proposed rules for deciding when to stop subdividing the groups; however, in practice, it may be preferable to produce a fairly exhaustive classification initially and decide the level of grouping later.

Ivimey-Cook and Proctor (1966) used the method of association analysis to classify stands of vegetation from salt marshes in western Ireland. They produced a classification into seven defined groups, and a residual eighth group in which all species used for separating the groups were absent (Table 5.2). The species used for group division are identified by capitals in the table. Another classification with reallocation of some stands is given in Section 5.3.

One attraction of monothetic methods is that they provide a simple binary key which can be used in the field for classifying further samples in terms of the presence or absence of a small number of species (Fig. 5.3). However, because each division depends on the state of just a single species, the final classification may not be very robust.

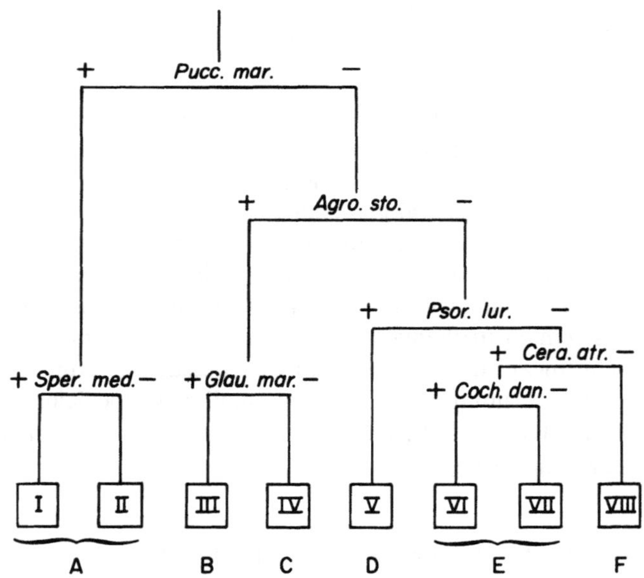

Fig. 5.3 A binary key for identifying types of salt-marsh habitat, derived from Ivimey-Cook and Proctor (1966). Division into eight groups is based on presence or absence of the species shown (for full names see Table 5.2). An additional grouping into six groups coded A to F is also marked.

An alternative method consists of dividing the samples according to their score on the principal axis of an ordination. This method is polythetic, i.e. group divisions are based on all species. Initially two groups are formed which are each subjected to a further ordination and division to form four groups, and so on. Although this method uses all species in classifying samples, it does not provide a simple key, suitable for field application, which would allow new samples to be assigned to their appropriate class. Hill, Bunce and Shaw (1975) developed a hybrid of these two methods which they termed 'indicator species analysis'. Essentially this involves selecting five 'indicator' species which closely reproduce the initial one-dimensional ordination. Each sample is then scored for presence/absence of each of these species, leading to an overall score on a six-point scale which is divided to form two groups. Each group is then further divided by carrying out another ordination and defining a new set of indicator species. The number of species used as indicators is somewhat arbitrary: five was chosen as a compromise between using a single species, for simplicity of field application, and using the complete ordinations from all species.

More recently Hill (1979b) has extended and modified his method to provide an associated classification of species based on the sample classification. Hill called the method 'two-way indicator species analysis' but admitted this name was somewhat misleading as much less emphasis is now placed on the final selection of, and classification by, indicator species. Hill's alternative name, 'dichotomized ordination analysis', is more appropriate as it emphasizes the link with ordination; this allows the hierarchy of classes to be ordered in an informative way. Examples of the use of the method are given by Kershaw and Looney (1985).

5.3 NON-HIERARCHICAL CLASSIFICATION

Often there is no particular advantage in the groups of species or sites being arranged in a hierarchy. A non-hierarchical method of classification, where groups have no joint structure, may then be preferred. These methods have two main aspects: the first is that they seek to partition the units into a specified number of groups to optimize some criterion; the second is the method by which the optimum is achieved. We will start with a discussion of optimization methods and leave the criteria until later.

Obviously with anything but very small numbers of units and/or required groups it is not feasible to try every possible arrangement (e.g. there are over 10,000 different ways of forming three groups from just ten units). In practice, some starting classification is chosen and then a series of possible moves of units among groups is tried; the move (if any) that best improves the criterion is made, and the process is repeated until no move gives an improvement. This is then taken to be the optimum grouping.

An initial classification may be obtained in many ways. Obviously one can

Table 5.2 **Presence (+)/absence (.) matrix of species in 70 samples from salt-marshes in western Ireland (Ivimey-Cook and Proctor, 1966). Samples are classified into eight groups by association analysis based on the occurrence of the seven species in capitals.**

	Samples																													
	I														II										III					
Species	1	2	3	4	5	6	7	8	9	10	11	12	13	14	15	16	17	18	19	20	21	22	23	24	25	26	27	28	29	30
COCHLEARIA DANCIA	.	.	.	.	.	.	.	.	.	.	.	.	.	.	.	+	+	.	.	.	.	.	.	.	+	.	+	.	.	.
Festuca rubra	.	+	.	.	.	.	.	.	.	.	+	.	.	+	+	+	+	+	.	+	+	+	.	.	+	+	+	+	+	+
Plantago coronopus	.	.	.	.	.	.	.	.	.	.	.	.	.	.	.	+	+	+	+	+	.	+	.	.	+	.	+	+	+	+
Armeria maritima	+	+	+	+	+	+	+	+	+	+	+	+	+	+	+	+	+	+	+	+	+	+	.	+	+	+	+	+	+	+
Plantago maritima	+	+	+	+	+	+	.	+	+	+	+	.	.	+	+	+	+	+	+	.	.	+	+	.	+	+	+	+	+	+
PUCCINELLIA MARITIMA	+	+	+	+	+	+	+	+	+	+	+	+	+	+	+	+	+	+	+	+	+	+	+	+	.	.	.	.	.	.
Triglochin maritima	.	.	+	.	.	.	.	.	.	.	.	.	+	.	.	.	.	.	.	.	.	.	.	.	.	.	.	.	+	.
GLAUX MARITIMA	+	+	+	+	+	+	.	.	+	+	.	.	.	+	+	.	.	.	+	+	.	+	.	+	+	+	+	+	+	+
Aster tripolium	+	+	+	+	+	+	+	+	+	+	+	+	.	.	+	.	.	.	.	+	+	+	+	+	.	.	.	.	.	.
Limonium humile	.	.	+	+	+	+	+	+	+	.	+	+	+	.	.	.	.	.	.	.	.	.	.	.	.	.	.	.	.	.
Salicornia europaea	.	+	+	+	+	+	+	+	+	+	.	.	+	.	.	.	.	.	.	.	.	.	.	.	.	.	.	.	.	.
SPERGULARIA MEDIA	+	+	+	+	+	+	+	+	+	+	+	+	+	+	.	.	.	.	.	.	.	.	.	.	.	.	.	.	.	.
Suaeda maritima	+	.	+	+	+	+	+	+	+	.	+	+	.	+	.	.	.	.	.	.	.	+	+	+	.	.	.	.	.	.
AGROSTIS STOLONIFERA	.	+	.	.	.	.	.	.	.	.	.	.	.	.	.	.	.	.	.	.	.	.	.	.	+	+	+	+	+	+
Carex distans	.	.	.	.	.	.	.	.	.	.	.	.	.	.	.	.	.	.	.	.	.	.	.	.	+	+	+	.	.	+
Trifolium repens	.	.	.	.	.	.	.	.	.	.	.	.	.	.	.	.	.	.	.	.	.	.	.	.	+	+	+	+	.	+
CERASTIUM ATROVIRENS	.	.	.	.	.	.	.	.	.	.	.	.	.	.	.	.	.	.	.	.	.	.	.	.	+	.	.	.	.	.
Tortella flavovirens	.	.	.	.	.	.	.	.	.	.	.	.	.	.	.	.	.	.	.	.	.	.	.	.	+	.	+	.	.	.
Leontodon autumnalis	.	.	.	.	.	.	.	.	.	.	.	.	.	.	.	.	.	.	.	.	.	.	.	.	+	+	+	+	.	.
Lotus corniculatus	.	.	.	.	.	.	.	.	.	.	.	.	.	.	.	.	.	.	.	.	.	.	.	.	+	.	.	.	.	.
Cerastium holosteoides	.	.	.	.	.	.	.	.	.	.	.	.	.	.	.	.	.	.	.	.	.	.	.	.	+	+	.	.	.	.
Catapodium marinum	.	.	.	.	.	.	.	.	.	.	.	.	.	.	.	.	.	.	.	.	.	.	.	.	.	.	.	.	.	.
Sagina nodosa	.	.	.	.	.	.	.	.	.	.	.	.	.	.	.	.	.	.	.	.	.	.	.	.	.	.	.	.	.	.
Amblystegium serpens	.	.	.	.	.	.	.	.	.	.	.	.	.	.	.	.	.	.	.	.	.	.	.	.	.	.	.	.	.	+
Cladonia pyxidata	.	.	.	.	.	.	.	.	.	.	.	.	.	.	.	.	.	.	.	.	.	.	.	.	.	.	.	.	.	.
Trichostomum brachydontium	.	.	.	.	.	.	.	.	.	.	.	.	.	.	.	.	.	.	.	.	.	.	.	.	.	.	.	.	.	.
Crithmum maritimum	.	.	.	.	.	.	.	.	.	.	.	.	.	.	.	.	.	.	.	.	+	.	.	.	.	.	.	.	.	.
Silene maritima	.	.	.	.	.	.	.	.	.	.	.	.	.	.	.	.	.	.	.	.	.	.	.	.	.	.	.	.	.	.
Pottia heimii	.	.	.	.	.	.	.	.	.	.	.	.	.	.	.	.	.	.	.	.	.	.	.	.	.	.	+	.	.	.
Atriplex hastata s.l.	.	.	.	.	.	.	.	.	.	.	+	.	.	.	.	.	+	.	.	.	+	.	+	.	.	.	.	.	.	.
Juncus gerardii	.	.	.	.	.	.	.	.	.	.	.	.	.	.	.	.	.	.	.	.	.	.	.	.	.	.	.	.	+	+
Sagina maritima	.	.	.	.	.	.	.	.	.	.	.	.	.	.	.	.	.	.	.	.	.	.	.	.	.	.	.	.	.	.
Barbula fallax	.	.	.	.	.	.	.	.	.	.	.	.	.	.	.	.	.	.	.	.	.	.	.	.	.	.	.	.	.	.
PSORA LURIDA	.	.	.	.	.	.	.	.	.	.	.	.	.	.	.	.	.	.	.	.	.	.	.	.	.	.	.	.	.	.
Anthyllis vulneraria	.	.	.	.	.	.	.	.	.	.	.	.	.	.	.	.	.	.	.	.	.	.	.	.	.	.	.	.	.	.
Bellis perennis	.	.	.	.	.	.	.	.	.	.	.	.	.	.	.	.	.	.	.	.	.	.	.	.	.	.	.	.	.	.
Camptothecium lutescens	.	.	.	.	.	.	.	.	.	.	.	.	.	.	.	.	.	.	.	.	.	.	.	.	.	.	.	.	.	.
Centaurium erythraea	.	.	.	.	.	.	.	.	.	.	.	.	.	.	.	.	.	.	.	.	.	.	.	.	.	.	.	.	.	.
Euphrasia nemorosa	.	.	.	.	.	.	.	.	.	.	.	.	.	.	.	.	.	.	.	.	.	.	.	.	.	.	.	.	.	.
Koeleria cristata	.	.	.	.	.	.	.	.	.	.	.	.	.	.	.	.	.	.	.	.	.	.	.	.	.	.	.	.	.	.
Linum catharticum	.	.	.	.	.	.	.	.	.	.	.	.	.	.	.	.	.	.	.	.	.	.	.	.	.	.	.	.	.	.
Ceratodon purpureus	.	.	.	.	.	.	.	.	.	.	.	.	.	.	.	.	.	.	.	.	.	.	.	.	.	.	.	.	.	.
Artemisia maritima	.	.	.	.	+	.	.	.	.	.	+	.	.	.	.	.	.	.	.	.	.	.	+	+	.	.	.	.	.	.
Carex extensa	.	.	.	.	.	.	.	.	.	.	.	.	.	.	.	.	.	.	.	+	+	.	.	.	.	.	.	.	+	.
Cochlearia anglica	.	.	.	.	.	.	.	+	.	.	.	.	.	+	.	.	.	.	.	.	.	+	.	.	.	.	.	.	.	.
Juncus maritimus	.	.	.	.	.	.	.	.	.	.	.	.	.	.	.	.	.	.	.	.	.	.	.	.	.	.	.	.	.	.
Limonium transwallianum	.	.	.	.	.	.	.	.	.	.	.	.	.	.	.	.	+	+	+	+	.	.	.	.	.	.	.	.	.	.
Parapholis strigosa	+	+	.	.	.	.	.	.	.	.	.	.	.	.	.	.	.	.	.	.	.	.	.	.	.	.	.	.	.	.
Polygala vulgaris	.	.	.	.	.	.	.	.	.	.	.	.	.	.	.	.	.	.	.	.	.	.	.	.	.	.	.	.	.	.
Sonchus sp.	.	.	.	.	.	.	.	.	.	.	.	.	.	.	.	.	.	.	.	.	.	.	.	.	.	.	.	.	.	.
Spergularia rupicola	.	.	.	.	.	.	.	.	.	.	.	.	.	.	.	.	+	+	.	.	.	.	.	.	.	.	.	.	.	.
Tripleurospermum maritimum	.	.	.	.	.	.	.	.	.	.	.	.	.	.	.	.	.	.	.	.	.	.	.	.	.	.	.	.	.	.
Pleurochaete squarrosa	.	.	.	.	.	.	.	.	.	.	.	.	.	.	.	.	.	.	.	.	.	.	.	.	.	.	.	.	.	.
Tortella nitida	.	.	.	.	.	.	.	.	.	.	.	.	.	.	.	.	.	.	.	.	.	.	.	.	.	.	.	.	.	.

					IV						V			VI					VII												VIII								
31	32	33	34	35	36	37	38	39	40	41	42	43	44	45	46	47	48	49	50	51	52	53	54	55	56	57	58	59	60	61	62	63	64	65	66	67	68	69	70
+	+	.	+	.	.	+	.	.	.	.	.	.	.	+	+	+	+	+	.	.	.	.	.	.	.	.	.	.	.	.	.	.	.	.	+	+	.	+	+
+	+	+	.	+	+	+	+	.	+	+	.	.	.	+	+	+	+	+	+	+	+	+	+	+	+	+	+	.	+	+	+	.	.	.	.	.	+	.	+
.	.	+	.	+	+	+	+	+	+	+	+	+	+	+	+	+	+	+	+	+	+	+	+	+	+	+	+	.	.	+	+	.	.	+	+	.	.	.	+
+	+	+	+	+	+	+	+	+	+	+	+	+	.	+	+	+	+	+	+	+	+	+	+	+	+	+	+	+	+	+	+	+	.	+	+	+	+	+	+
+	+	+	+	+	+	.	+	+	+	+	+	+	+	.	.	.	.	.	.	.	.	.	.	.	+	+	.	.	.	.	.	.	.	.	.	.	.	.	.
.	.	.	.	.	.	.	.	.	.	.	.	.	.	.	.	.	.	.	.	.	.	.	.	.	.	.	.	.	.	.	.	.	.	.	.	.	.	.	.
+	+	+	+	.	.	.	.	.	.	.	.	.	.	.	.	.	.	.	.	.	.	.	.	.	.	.	.	.	.	.	.	.	.	.	.	.	.	.	.
+	+	+	+	+	.	.	.	.	.	.	.	.	.	.	.	.	.	.	.	.	.	.	.	.	.	.	.	.	.	.	.	.	.	.	.	.	.	.	.
+	.	.	+	.	.	.	.	.	.	.	.	.	.	.	.	+	.	+	.	.	.	.	.	.	.	.	.	.	+	.	+	+	.	+	+	+	+	.	.
.	.	.	.	.	.	.	.	.	.	.	.	.	.	.	.	.	.	.	.	.	.	.	.	.	.	.	.	.	.	.	.	.	.	.	.	.	.	.	.
.	.	.	.	.	.	.	.	.	.	.	.	.	.	.	.	.	.	.	.	.	.	.	.	.	.	.	.	.	.	.	.	.	.	.	.	.	.	.	.
.	.	.	.	+	.	.	.	.	.	.	.	.	.	.	.	.	.	.	.	.	.	.	.	.	.	.	.	.	.	.	.	.	.	.	.	.	.	.	.
.	.	.	.	.	.	.	.	.	.	.	.	.	.	.	.	.	.	.	.	.	.	.	.	.	.	.	.	.	.	.	.	+	.	.	.	.	.	.	.
+	+	+	+	+	+	+	+	+	+	+	.	.	.	.	.	.	.	.	.	.	.	.	.	.	.	.	.	.	.	.	.	.	.	.	.	.	.	.	.
.	.	+	.	.	+	+	+	+	+	+	.	.	.	+	.	.	.	.	.	.	.	.	.	.	.	.	.	.	.	.	.	.	.	.	.	.	.	.	.
.	.	+	.	+	+	+	+	+	+	+	.	+	.	.	.	.	.	.	.	.	+	+	.	.	.	.	.	.	.	+	.	.	.	.	.	.	.	.	.
.	.	.	.	.	.	+	+	+	.	.	.	.	.	+	+	+	+	+	+	+	+	+	+	+	+	+	+	+	+	+	.	.	.	.	.	.	.	.	.
.	.	.	.	.	+	.	.	.	+	+	.	.	.	+	+	.	.	.	+	.	+	+	+	+	+	+	+	+	+	.	.	.	.	+	+	.	.	.	+
.	+	+	+	.	.	.	.	.	.	+	.	.	.	+	.	.	.	.	.	.	.	+	.	.	.	.	.	.	+	+	.	.	.	.	.	.	.	.	.
.	.	.	.	.	+	+	+	.	.	.	.	.	.	.	+	.	.	.	.	.	.	.	.	.	+	.	+	.	.	.	.	.	.	.	.	.	.	.	+
.	.	.	.	.	+	+	+	.	.	+	.	+	+	.	.	.	.	.	.	.	+	.	.	.	.	.	.	.	.	+	+	.	.	.	.	.	.	.	.
.	.	.	.	.	.	.	.	.	.	.	+	+	+	+	.	.	.	.	.	.	.	.	.	.	.	.	.	+	.	+	.	.	.	.	.	.	.	+	+
.	.	.	.	.	+	+	+	.	.	.	+	+	+	+	+	.	.	.	.	.	.	.	.	.	.	.	.	.	.	.	.	.	.	.	.	.	.	.	.
.	.	.	.	.	.	.	+	.	+	+	+	+	+	+	.	.	.	.	+	.	+	.	.	+	.	.	.	.	.	.	.	.	.	.	.	.	.	.	.
.	.	.	.	.	.	.	.	.	.	.	+	.	+	+	.	.	+	+	.	.	.	.	.	+	.	.	.	.	.	.	.	.	.	.	.	.	.	.	.
.	.	.	.	.	.	.	.	.	+	.	+	+	+	+	+	.	.	.	.	.	.	.	.	.	.	.	.	.	.	.	.	.	.	.	.	.	.	+	+
.	.	.	.	.	.	.	.	.	.	+	.	.	.	+	.	+	+	.	.	.	.	.	.	+	.	.	.	+	+	.	+	.	.	.	+	.	+	.	.
.	.	.	.	.	+	.	.	.	.	.	.	+	.	+	.	+	.	.	+	.	+	.	+	.	+	+	+	+	.	+	+	.	.	.	+	.	.	+	.
.	.	.	.	.	.	.	.	.	.	.	.	.	.	+	.	.	+	+	.	.	.	.	.	.	.	.	.	.	.	.	.	.	.	.	.	.	.	.	.
.	.	.	+	.	.	.	.	.	.	.	.	.	.	.	.	.	.	.	.	.	.	.	.	.	.	.	.	.	.	.	+	+	+	.	.	.	.	.	.
+	+	+	+	+	.	.	.	.	.	.	.	.	.	.	.	.	.	.	.	.	.	.	.	.	.	.	.	.	.	.	.	.	+	.	.	.	.	.	.
.	.	.	.	.	.	.	.	.	.	.	+	+	+	.	.	.	.	+	.	.	+	.	.	+	.	.	.	.	.	.	.	.	.	.	.	.	.	.	.
.	.	.	.	.	.	.	.	.	.	.	+	+	.	.	.	.	.	.	.	.	.	.	.	.	.	.	.	.	.	.	.	.	.	.	.	.	.	.	.
.	.	.	.	.	.	.	.	.	.	.	+	+	+	.	.	.	.	.	.	.	.	.	.	.	.	.	.	.	.	.	.	.	.	.	.	.	.	.	.
.	.	.	.	.	+	.	.	.	.	+	.	+	.	.	.	.	.	.	.	.	.	.	.	.	.	.	.	.	.	.	.	.	.	.	.	.	.	.	.
.	.	.	.	.	.	.	.	.	.	.	.	+	+	.	.	.	.	.	.	.	.	.	.	.	.	.	.	.	.	.	.	.	.	.	.	.	.	.	.
.	.	.	.	.	.	.	.	.	.	.	.	+	+	.	.	.	.	.	.	.	.	.	.	.	.	.	.	.	.	+	.	.	.	.	.	.	.	.	.
.	.	.	.	.	+	.	+	.	.	.	.	.	.	.	.	.	.	.	.	.	.	.	.	.	.	.	.	.	.	.	.	.	.	.	.	.	.	.	.
.	.	.	.	.	+	+	+	.	.	.	.	.	.	.	.	.	.	.	.	.	.	.	.	.	.	.	.	.	.	.	.	.	.	.	.	.	.	.	.
.	.	.	.	.	.	+	+	.	.	.	.	.	.	.	.	.	.	.	.	.	.	.	.	.	.	.	.	.	.	.	.	.	.	.	.	.	.	.	.
.	.	.	.	.	+	+	+	.	.	.	.	.	.	.	.	.	.	.	.	.	.	.	.	.	.	.	.	.	.	.	.	.	.	.	.	.	.	.	.
.	.	.	.	+	.	.	+	.	+	.	.	.	.	.	.	.	.	.	.	.	.	.	.	.	.	.	.	.	.	.	.	.	.	.	.	.	.	.	.
.	.	.	.	.	.	.	.	.	.	.	.	.	.	.	.	.	.	.	.	.	.	.	.	.	.	.	.	.	.	.	.	.	.	.	.	.	.	.	.
+	+	.	.	.	.	.	.	.	.	.	.	.	.	.	.	.	.	.	.	.	.	.	.	.	.	.	.	.	.	.	.	.	.	.	.	.	.	.	.
.	.	.	.	.	.	.	.	.	.	.	.	.	.	.	.	.	.	.	.	.	.	.	.	.	.	.	.	.	.	.	.	.	.	.	.	.	.	.	.
.	+	.	+	.	.	.	.	.	.	.	.	.	.	.	.	.	.	.	.	.	.	.	.	.	.	.	.	.	.	.	.	.	.	.	.	.	.	.	.
.	.	.	.	.	.	.	.	.	.	.	.	.	.	.	.	.	+	.	.	.	.	.	.	.	.	.	.	.	.	.	.	.	.	.	.	.	.	.	.
.	.	.	.	.	.	.	.	.	.	.	.	.	.	.	.	.	.	.	.	.	.	.	.	.	.	.	.	.	.	.	.	.	.	.	.	.	.	.	.
.	.	.	.	.	+	+	.	.	.	.	.	.	.	.	.	.	.	.	.	.	.	.	.	.	.	.	.	.	.	.	.	.	.	.	.	.	.	.	.
.	.	.	.	.	.	.	.	.	.	.	.	.	.	.	.	.	.	.	.	.	.	.	.	.	.	.	.	.	.	.	.	.	.	.	.	.	.	+	+
.	.	.	.	.	.	.	.	.	.	.	.	.	.	.	.	.	.	.	.	.	.	.	.	.	.	.	.	.	.	.	+	+	.	.	.	+	.	.	.
.	.	.	.	.	.	.	.	.	.	.	.	.	.	.	.	.	.	.	.	.	.	.	.	.	.	.	+	.	.	.	+	.	.	.	.	.	+	.	.
.	.	.	.	.	.	.	.	.	.	.	.	.	.	.	.	.	+	.	.	.	.	.	.	.	.	.	+	.	.	.	.	.	.	.	.	.	.	.	.
.	.	.	.	.	.	.	.	.	.	.	.	.	.	.	.	.	.	.	.	.	.	.	.	.	.	.	.	.	.	.	.	.	.	.	.	.	.	+	+

make a random, or pseudo-random, allocation of the units to the groups and start from there. However, a bad initial grouping can provide an anomalous final solution at some local optimum; it is therefore more judicious to take a reasonably acceptable initial grouping, from say a hierarchical classification or a partitioning of the ordination space. Once an initial grouping is provided, computer programs vary in the types of move available and the order in which they are tried. The algorithm in Genstat has two types of move: transfers and swops. Transfers consist simply of moving one unit to a different group; swops involve the exchange of two units from different groups. Since swops are costlier computationally and may be achieved automatically by two transfers, Genstat starts by using transfers, taking the best at each stage, until no improvement can be made. Only then are swops considered: the best (if any) is made and the algorithm returns to assessing transfers. More elaborate types of move are obviously possible, as are different ways of using them.

However elaborate the moves become, nothing short of an exhaustive search can guarantee to find the global optimum. There may well be many local optima, often with very different allocations of the units to the groups, so it is advisable to rerun any analysis from a few different starting points. Obviously a good initial allocation will help; in the example of Section 5.5 we use two, both from agglomerative classifications.

Various criteria have been used for describing the optimum classification. Here we discuss two: minimum within-group sum of squares and maximal predictive classification.

The sum of squares method can be used for a classification of units which can be represented as points in Euclidean space with p dimensions. Suppose it is desired to classify n units with coordinates z_{ik} $(i=1, \ldots, n; k=1, \ldots, p)$ into g groups. Then the required classification is that which minimizes the total within-group sum of squares of units about their group centroids, i.e.

$$\sum_{m=1}^{g} \sum_{i} \sum_{k=1}^{p} (z_{ik} - \bar{z}_{mk})^2,$$

where the summation over i is only for those units in the mth group; this has centroid $(\bar{z}_{m1}, \ldots, \bar{z}_{mp})$. An alternative criterion is to maximize the total Mahalanobis distance between groups (Section 1.4.3).

Maximal predictive classification (Gower, 1974) operates only on binary data in the form of a units-by-variates matrix. Although developed primarily for taxonomic classification, this method can be used for classifying sites based on the presence/absence of species. Maximal predictive classification is based on defining, for each group, a predictor which describes the most common state of each variable for that group. Thus for the artificial data matrix of Table 5.3, the group predictor for units 1, 2, 3 and 4 is $(+, +, -, +, +)$. If this is used to predict the properties of unit 3, for example, it will correctly predict the states of four of the five variables, but incorrectly predict the state of variable 2.

Table 5.3 Group predictors for an artificial data set comprising eight units described by five binary variables, and optimal division into three groups.

Group	Unit	Variable 1	2	3	4	5	Number of correct predictions
1	1	+	+	−	−	+	4
	2	+	+	−	+	+	5
	3	+	−	−	+	+	4
	4	−	+	−	+	+	4
	Group predictor	(+	+	−	+	+)	
2	5	+	+	+	−	−	5
	Group predictor	(+	+	+	−	−)	
3	6	+	−	−	−	+	4
	7	−	+	−	−	+	4
	8	−	−	−	+	+	4
	Group predictor	(−	−	−	−	+)	
	Value of optimizing criterion						34

For group 2, which consists of a single unit, the group predictor is the same as the unit itself. For each arrangement of units into a specified number of groups, the total number of correct predictions is calculated and this criterion is maximized to give the final classification. As the method is usually defined, joint absences of a species for a unit and its group predictor contribute as much to the criterion as joint occurrences; in some circumstances it may be more appropriate to ignore joint absence. In common with monothetic divisive methods described in Section 5.2, maximal predictive classification provides a reduced list of key variables or species which can be used for field classification. Any new unit is compared with each group predictor and is assigned to that group with which it has the greatest number of matches (joint occurrences, or both joint occurrences and absences). For example, in Table 5.3 a unit characterized as (+, +, −, −, −) would be assigned to group 2.

5.3.1 Example

The method of maximal predictive classification is here applied to the salt marsh samples of Table 5.2. Ivimey-Cook and Proctor's (1966) association analysis produced eight groups, but we initially consider classification into only six groups, coded A to F: these are shown on Fig. 5.3 and combine groups I and II, to form group A, and VI and VII as E. On identifying the group predictors for this six-group classification, the number of correct predictions was calculated to be 3460. Applying the Genstat maximal predictive algorithm, a substantially different grouping was obtained which gave 3482 correct predictions: essentially this entailed combining the majority of units from group II with group VIII rather than with group I. Indeed, the algorithm provided 20 variants of this optimal six-group classification, each with the same number of correct predictions, which largely comprised different ways of dividing the combined group II + VI + VII + VIII. However, by experimenting with different initial groupings based on the ordination of Fig. 5.4, a marginally better classification, with 3483 correct predictions, was obtained: this appears to be a global maximum. Moreover, the classification is unique save for a slight indeterminancy over the separate allocation of samples 23 and 24: equal values of the criterion are achieved regardless of whether they are placed in group A or group F. Table 5.4 shows the reclassified data matrix with

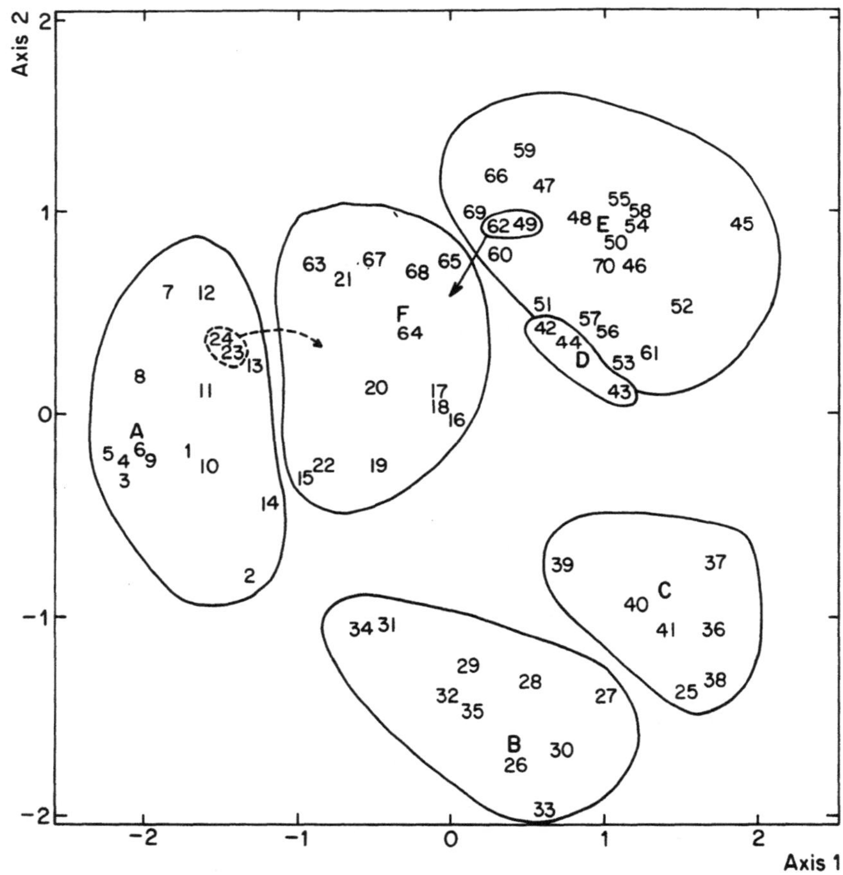

Fig. 5.4 Principal components ordination of the salt-marsh samples (Table 5.2) showing the six groups obtained by maximal predictive classification (Table 5.4). The locations of the group predictors are indicated by the letters A to F.

the group predictors indicated. Species not included as a positive predictor for any group are omitted – these include *Armeria maritima* which is found in every group and 67 of the 70 samples. Whereas groups A, B, C and D are identified by the presence of a number of species, groups E and F are very sparse and are identified by the presence of only four and five species respectively, none of which are unique to that group. Indeed, if a classification into five groups is preferred, the algorithm merges groups E and F, while the other groups remain largely unchanged. We would therefore question the justification for further splitting groups E and F, to make eight groups in all, as suggested by the analysis of Ivimey-Cook and Proctor (1966).

Figure 5.4 shows the grouping of samples on the two-dimensional ordination produced by principal components analysis; this is equivalent to PCO using the simple matching coefficient for similarity. Like maximal predictive classification, this gives full weight to joint absences and, somewhat surprisingly, was found to give a more acceptable ordination of samples than correspondence analysis. The scores for each group predictor may be calculated from the species loadings and added to the ordination plot: in Fig. 5.4 the predictors are seen to lie approximately at the centres of their respective groups. On the two-dimensional ordination, group D appears to be closely related to group E. However, these groups are widely separated in the third and higher dimensions, as might be deduced from their very different predictors shown in Table 5.4.

5.4 VISUAL DISPLAYS FOR CLASSIFICATION

Like other multivariate methods, the effectiveness of classification relies heavily upon visual displays of the results. In this section we describe some procedures for displaying classifications. We have already shown the usefulness of including information on the grouping of units on an ordination to highlight possible misinterpretations (see the examples of Section 5.3.1 and Section 5.5). Here we shall describe displays which relate directly to classification and are primarily associated with hierarchical methods.

5.4.1 Dendrograms

The standard display for hierarchical methods is the dendrogram: an example for the 20 commonest species on Park Grass is shown in Fig. 5.5. Here the individual units (species) are shown on the left-hand side, and units and clusters are successively merged as one moves across the display, from left to right. The display serves perfectly well for a divisive hierarchical method, in which case moving across the display from right to left shows the division of clusters into smaller clusters. The horizontal axis shows the similarity at which clusters are joined. The units may be divided into a number of distinct groups by drawing a single vertical line on the display: all clustering to the right of the

Table 5.4 Reclassification of samples in Table 5.2 into six groups by maximal predictive classificatio
Species identified in a group by bold type are those whose presence is part of the group predictor. Species which do not act as predictors for any group are omitted.

Species	A																B										C			
	1	2	3	4	5	6	7	8	9	10	11	12	13	14	23	24	26	27	28	29	30	31	32	33	34	35	25	36	37	38
Spergularia media	+	+	+	+	+	+	+	+	+	+	+	+	+	+	.	.	.	.	.	.	.	.	.	.	.	+	.	.	.	.
Suaeda maritima	+	.	+	+	+	+	+	+	+	.	+	+	.	+	+	+	.	.	.	.	.	.	.	.	.	.	.	.	.	.
Salicornia europaea	.	+	+	+	+	+	+	+	+	+	.	.	+	.	.	.	.	.	.	.	.	.	.	.	.	.	.	.	.	.
Limonium humile	.	.	+	+	+	+	+	+	+	.	+	+	+	.	.	.	.	.	.	.	.	.	.	.	.	.	.	.	.	.
Aster tripolium	+	+	+	+	+	+	+	+	+	+	+	+	.	.	+	+	.	.	.	.	.	+	.	.	+	.	.	.	.	.
Puccinellia maritima	+	+	+	+	+	+	+	+	+	+	+	+	+	+	+	+	.	.	.	.	.	.	.	.	.	.	.	.	.	.
Glaux maritima	+	+	+	+	+	+	.	.	+	+	.	.	.	+	.	+	+	+	+	+	+	+	+	+	+	+	+	.	.	.
Plantago maritima	+	+	+	+	+	+	.	+	+	+	+	.	.	+	+	.	+	+	+	+	+	+	+	+	+	+	+	+	.	+
Plantago coronopus	.	.	.	.	.	.	.	.	.	.	.	.	.	.	.	.	.	+	+	+	+	.	.	+	.	+	+	+	+	+
Festuca rubra	.	+	.	.	.	.	.	.	.	.	+	.	.	+	.	.	+	+	+	+	+	+	+	+	.	+	+	+	+	+
Triglochin maritima	.	.	+	.	.	.	.	.	.	.	.	.	+	.	.	.	.	.	.	+	.	+	+	+	+	.	.	.	.	.
Juncus gerardii	.	.	.	.	.	.	.	.	.	.	.	.	.	.	.	.	.	.	.	+	+	+	+	+	+	+	.	.	.	.
Leontodon autumnalis	.	.	.	.	.	.	.	.	.	.	.	.	.	.	.	.	+	+	+	.	.	.	+	+	+	.	+	.	.	.
Agrostis stolonifera	.	+	.	.	.	.	.	.	.	.	.	.	.	.	.	.	+	+	+	+	+	+	+	+	+	+	+	+	+	+
Trifolium repens	.	.	.	.	.	.	.	.	.	.	.	.	.	.	.	.	+	+	+	.	+	.	.	+	.	+	+	+	+	+
Carex distans	.	.	.	.	.	.	.	.	.	.	.	.	.	.	.	.	+	+	.	.	+	.	.	+	.	.	+	+	+	+
Lotus corniculatus	.	.	.	.	.	.	.	.	.	.	.	.	.	.	.	.	.	.	.	.	.	.	.	.	.	.	+	+	+	+
Cerastium holosteoides	.	.	.	.	.	.	.	.	.	.	.	.	.	.	.	.	+	.	.	.	.	.	.	.	.		+	+	+	+
Tortella flavovirens	.	.	.	.	.	.	.	.	.	.	.	.	.	.	.	.	.	+	.	.	.	.	.	.	.	.	+	+	.	.
Cerastium atrovirens	.	.	.	.	.	.	.	.	.	.	.	.	.	.	.	.	.	.	.	.	.	.	.	.	.	.	+	.	+	+
Camptothecium lutescens	.	.	.	.	.	.	.	.	.	.	.	.	.	.	.	.	.	.	.	.	.	.	.	.	.	.	.	.	.	.
Bellis perennis	.	.	.	.	.	.	.	.	.	.	.	.	.	.	.	.	.	.	.	.	.	.	.	.	.	.	.	.	.	.
Psora lurida	.	.	.	.	.	.	.	.	.	.	.	.	.	.	.	.	.	.	.	.	.	.	.	.	.	.	.	.	.	.
Barbula fallax	.	.	.	.	.	.	.	.	.	.	.	.	.	.	.	.	.	.	.	.	.	.	.	.	.	.	.	.	.	.
Sagina maritima	.	.	.	.	.	.	.	.	.	.	.	.	.	.	.	.	.	.	.	.	.	.	.	.	.	.	.	.	.	.
Catapodium marinum	.	.	.	.	.	.	.	.	.	.	.	.	.	.	.	.	.	.	.	.	.	.	.	.	.	.	.	.	.	.
Sagina nodosa	.	.	.	.	.	.	.	.	.	.	.	.	.	.	.	.	.	.	.	.	.	.	.	.	.	.	.	+	+	+
Amblystegium serpens	.	.	.	.	.	.	.	.	.	.	.	.	.	.	.	.	.	.	.	.	+	.	.	.	.	.	.	.	.	+
Cladonia pyxidata	.	.	.	.	.	.	.	.	.	.	.	.	.	.	.	.	.	.	.	.	.	.	.	.	.	.	.	.	.	.
Trichostomum brachydontium	.	.	.	.	.	.	.	.	.	.	.	.	.	.	.	.	.	.	.	.	.	.	.	.	.	.	.	.	.	.
Silene maritima	.	.	.	.	.	.	.	.	.	.	.	.	.	.	.	.	.	.	.	.	.	.	.	.	.	.	.	+	.	.

line is ignored and the groups are defined by the clustering to the left of the line. In practice, however, clusters may be obtained which are more easily interpretable by cutting certain joins on the dendrogram rather than choosing a single similarity threshold. Thus in Fig. 5.5 one might assign the species to four groups:

(1) *Leontodon*, *Poterium*, *Lolium*, *Trifolium*, *Ranunculus*, *Plantago*, *Helictotrichon*;
(2) *Anthriscus*, *Taraxacum*, *Heracleum*, *Lathyrus*, *Dactylis*, *Poa trivialis*;
(3) *Alopecurus*, *Arrhenatherum*;
(4) *Agrostis*, *Anthoxanthum* with possibly *Festuca*;
Poa pratensis and *Holcus* are not associated with any species group.

The dendrogram of Fig. 5.5 may alternatively be displayed with axes transposed so that the individual units occur at the top or bottom of the

Samples																																						
D					E																			F														
40	41	42	43	44	45	46	47	48	50	51	52	53	54	55	56	57	58	59	60	61	66	69	70	15	16	17	18	19	20	21	22	49	62	63	64	65	67	68
.	.	.	.	.	.	.	.	.	.	.	.	.	.	.	.	.	.	.	.	.	.	.	.	.	.	.	.	.	.	.	.	.	.	.	.	.	.	.
.	.	.	.	.	.	.	.	.	.	.	.	.	.	.	.	.	.	.	.	.	.	.	.	.	.	.	.	.	.	.	+	.	.	+	.	.	.	.
.	.	.	.	.	.	.	.	.	.	.	.	.	.	.	.	.	.	.	.	.	.	.	.	.	.	.	.	.	.	.	.	.	.	.	.	.	.	.
.	.	.	.	.	.	.	.	.	.	.	.	.	.	.	.	.	.	.	.	.	.	.	.	.	.	.	.	.	.	.	.	.	.	.	.	.	.	.
.	.	.	.	.	.	.	+	.	.	.	.	.	.	.	.	.	.	.	+	.	+	.	.	+	.	.	.	.	+	+	+	+	+	+	.	+	+	+
.	.	.	.	.	.	.	.	.	.	.	.	.	.	.	.	.	.	.	.	.	.	.	.	+	+	+	+	+	+	+	+	.	.	.	.	.	.	.
.	.	.	.	.	.	.	.	.	.	.	.	.	.	.	.	.	.	.	.	.	.	.	.	+	.	.	.	+	+	.	+	.	.	.	.	.	.	.
+	+	+	+	+	.	.	.	.	.	.	.	.	.	.	+	+	.	.	.	.	.	.	.	+	+	+	+	+	.	.	+	.	.	.	.	.	.	.
+	+	+	+	+	+	+	+	+	+	+	+	+	+	+	+	+	+	.	.	+	+	.	+	.	+	+	+	+	+	.	+	+	+	.	.	+	.	.
+	+	.	.	.	+	+	+	+	+	+	+	+	+	+	+	+	+	.	+	+	.	.	+	+	+	+	+	.	+	+	+	+	+	.	.	.	.	+
.	.	.	.	.	.	.	.	.	.	.	.	.	.	.	.	.	.	.	.	.	.	.	.	.	.	.	.	.	.	.	.	.	.	.	.	.	.	.
.	.	.	.	.	.	.	.	.	.	.	.	.	.	.	.	.	.	.	.	.	.	.	.	.	.	.	.	.	.	.	.	.	.	.	+	.	.	.
.	+	.	.	.	+	.	.	.	.	.	.	+	.	.	.	.	.	.	+	+	.	.	.	.	.	.	.	.	.	.	.	.	.	.	.	.	.	.
+	+	.	.	.	.	.	.	.	.	.	.	.	.	.	.	.	.	.	.	.	.	.	.	.	.	.	.	.	.	.	.	.	.	.	.	.	.	.
+	+	.	+	.	.	.	.	.	.	.	+	+	.	.	.	.	.	.	.	+	.	.	.	.	.	.	.	.	.	.	.	.	.	.	.	.	.	.
+	+	.	.	.	+	.	.	.	.	.	.	.	.	.	.	.	.	.	.	.	.	.	.	.	.	.	.	.	.	.	.	.	.	.	.	.	.	.
.	.	.	.	.	.	+	.	.	.	.	.	.	.	.	+	.	+	.	.	.	.	.	+	.	.	.	.	.	.	.	.	.	.	.	.	.	.	.
.	+	.	+	+	.	.	.	.	.	.	+	.	.	.	.	.	.	.	.	+	.	.	.	.	.	.	.	.	.	.	.	.	+	.	.	.	.	.
+	+	.	.	.	+	+	.	.	+	.	+	+	+	+	+	+	+	+	+	.	+	.	+	.	.	.	.	.	.	.	.	.	.	.	.	+	.	.
.	.	.	.	.	+	+	+	+	+	+	+	+	+	+	+	+	+	+	+	+	.	.	.	.	.	.	.	.	.	.	.	+	.	.	.	.	.	.
.	.	.	+	+	.	.	.	.	.	.	.	.	.	.	.	.	.	.	.	+	.	.	.	.	.	.	.	.	.	.	.	.	.	.	.	.	.	.
.	.	.	+	+	.	.	.	.	.	.	.	.	.	.	.	.	.	.	.	.	.	.	.	.	.	.	.	.	.	.	.	.	.	.	.	.	.	.
.	.	+	+	+	.	.	.	.	.	.	.	.	.	.	.	.	.	.	.	.	.	.	.	.	.	.	.	.	.	.	.	.	.	.	.	.	.	.
.	.	+	+	.	.	.	.	.	.	.	.	.	.	.	.	.	.	.	.	.	.	.	.	.	.	.	.	.	.	.	.	.	.	.	.	.	.	.
.	.	+	+	+	.	.	.	.	.	.	+	.	.	+	.	.	.	.	.	.	.	.	.	.	.	.	.	.	.	.	.	+	.	.	.	.	.	.
.	.	+	+	+	+	.	.	.	.	.	.	.	.	.	.	.	.	+	.	+	.	+	+	.	.	.	.	.	.	.	.	.	.	.	.	.	.	.
.	.	+	+	+	+	+	.	.	.	.	.	.	.	.	.	.	.	.	.	.	.	.	.	.	.	.	.	.	.	.	.	.	.	.	.	.	.	.
+	+	+	+	+	+	.	.	.	+	.	+	.	.	+	.	.	.	.	.	.	.	.	.	.	.	.	.	.	.	.	.	.	.	.	.	.	.	.
.	.	+	.	+	+	.	.	+	.	.	.	.	.	+	.	.	.	.	.	.	.	.	.	.	.	.	.	.	.	.	.	+	.	.	.	.	.	.
+	.	+	+	+	+	+	.	.	.	.	.	.	.	.	.	.	.	.	.	.	.	+	+	.	.	.	.	.	.	.	.	.	.	.	.	.	.	.
.	.	.	+	.	+	.	+	.	+	.	+	.	+	.	+	+	+	+	.	+	+	+	.	.	.	.	.	.	.	.	.	.	+	.	.	.	.	.

display. In the former case the display can be viewed as a tree with a single trunk, spreading to various branches and leaves (the individuals); the latter looks more like a tree's root system, or a child's mobile.

Although the ordering of the units in the dendrogram must be partially constrained by the clustering, there is considerable latitude over the choice of a specific order. At every join of two groups in Fig. 5.5, either group may be placed above the other. Thus, returning to our analogy of the child's mobile (an idea originally due to John Gower), any arm of the mobile may be swung through 180° to result in a different dendrogram while showing the same clustering. Once this freedom is recognized, the descriptive power of the dendrogram may often be improved by choosing a different ordering of units. In Fig. 5.5, the units have been arranged so that when two groups merge the larger is always placed above the smaller; this ordering is particularly useful to show up chaining (Section 5.1). Here a mild degree of chaining is shown: after

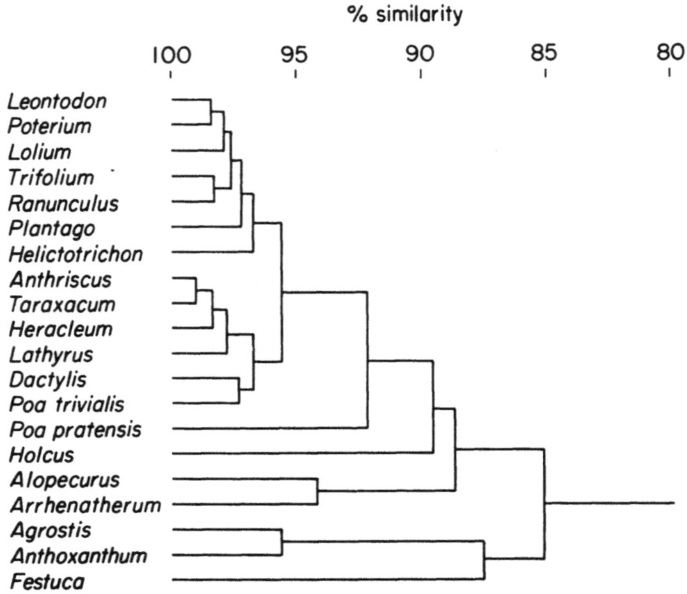

Fig. 5.5 A dendrogram of the major Park Grass species derived by centroid clustering. Inter-species similarities are based on the log abundances on the plots using the Euclidean measure of distance.

the merging of the two groups containing *Leontodon* and *Anthriscus*, the remaining two groups and the two disparate species are successively joined to this major cluster (see Fig. 3.5 where the major cluster occupies the centre of the ordination space). Here the dendrogram was obtained by centroid clustering: a more pronounced chaining would be expected from single-linkage clustering. An alternative ordering scheme would arrange the units so that they follow, as far as possible, some ordination: for example, this procedure is adopted in two-way indicator species analysis (Section 5.2).

5.4.2 Shade diagrams

Dendrograms are useful to show overall groupings but they mask individual similarities. Thus they may show one individual as belonging to a group when it is quite different from many of the individuals in the group and owes its inclusion simply to its closeness to one other individual in the group; this is especially common with single-linkage cluster analysis. Such anomalies can only be discovered by an inspection of the complete similarity matrix in which the individual units are reordered to match the order of units from the dendrogram. A more immediate visual impression of the inter-unit similarities is achieved from a shaded similarity matrix, where, instead of printing numerical values, a shaded box is drawn for each similarity value using progressively darker shading as the similarity increases (Section 2.1). This can work exceptionally well when the units are properly ordered, as for the artificial similarity matrix in Fig. 5.6. Then small groups (e.g. units 4, 5, 6, 7 and 8) show up as blocks of heavy shading on and near the diagonal of the matrix, atypical members of a group (unit 5) are shown by strips of lighter shading within darker blocks of shading, while individuals that are crucial to the merging of clusters (unit 7) are indicated by relatively darker shading within the block of between-group shading. The visual impact of these displays depends greatly on the number of different shades used and the cut-off values for the shading: often these decisions can be made more easily by first inspecting a histogram of the similarity values.

If the clustering is effective, the shaded similarity matrix should have dark blocks on its diagonal with lighter blocks appearing more frequently towards the bottom left-hand corner of the matrix. This effect can often be improved, for a given clustering, by rearranging units within the constraint imposed by the dendrogram, as described earlier. This is another example of how the final representation of a classification is often achieved by successive readjustment: here the initial dendrogram is used to produce the reordered similarity matrix, which is itself inspected to improve the ordering of units on the dendrogram.

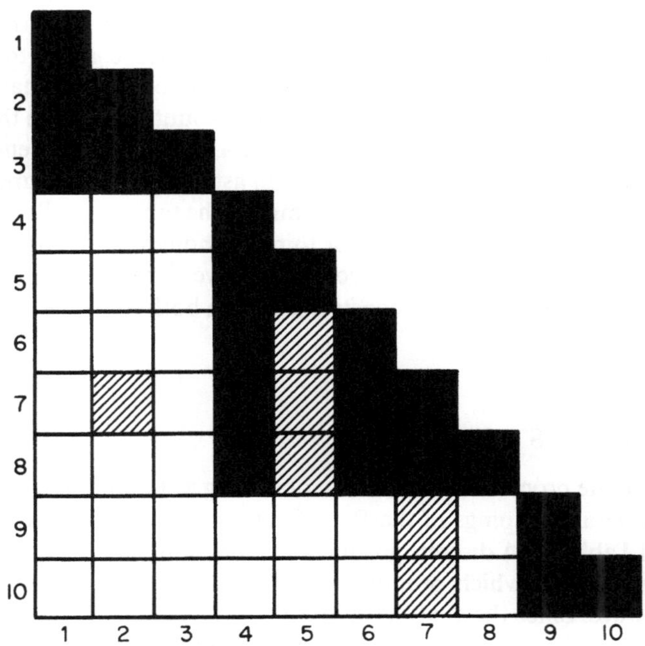

Fig. 5.6 Artificial similarity matrix for ten units showing its representation as a shaded matrix.

5.4.3 Combining displays

Obviously dendrograms and shade diagrams complement each other: the former indicates overall groupings, the latter shows individual similarities. Once a final ordering of units has been decided, a particularly useful type of display is a composite of these two: the easiest way to arrange this is to have the dendrogram, in the form of a tree, hanging below the shade diagram. Figure 5.7 displays the shaded similarity matrix for the Park Grass species with the associated dendrogram. The species have here been rearranged from the order of Fig. 5.5 so that the darker blocks in the similarity matrix lie closest to the diagonal. This new ordering is very similar to that on the principal ordination axis of Fig. 3.5. Inspection of the similarity matrix shows that the two central groups of species, containing *Leontodon* and *Anthriscus* (see Fig. 5.5), have a large degree of overlap. In particular, *Lolium* plays an important role in merging these two clusters, and the more isolated species to this central cluster.

When a similarity matrix has been constructed from a two-way data matrix, e.g. sites by species, the technique suggested above can be extended to a composite display of three components by displaying the data matrix to the left of the shade diagram. Obviously the rows of the matrix, i.e. for the items being clustered, must be reordered to match those of the shade diagram and hence of the dendrogram. As with the similarity matrix, it may be preferable to replace the actual data values by a display that is easier to scan: this may be done by transforming the variables to a common range, e.g. one to ten, thus retaining a numerical display; alternatively, symbols or shading may be used. For example, with abundance data in the form of counts roughly in the range 0 to 100, three symbols might be used: blank, or a full-stop, for absence; a plus sign for counts in the range one to ten; and an asterisk for counts greater than ten. It may be beneficial to order the columns of the table, as well as the rows. Table 3.2 gives an idea of this sort of thing, although it was obtained in a somewhat different way. A very comprehensive form of display, shown schematically in Fig. 5.8, can be constructed when both rows and columns are clustered.

5.5 CASE STUDY

To illustrate the procedures involved in obtaining a suitable classification, we consider several groupings of the Park Grass plots based on the vegetation samples of Table 1.1. A discrete grouping of plots is useful for indicating those fertilizer treatments which produce a similar composition of species in the meadow sward. Since the analysis of Section 3.2.2 identified two gradients (of soil pH and total plot productivity) which were primarily responsible for the species distribution over the plots, classification was initially aimed at producing four groups of plots.

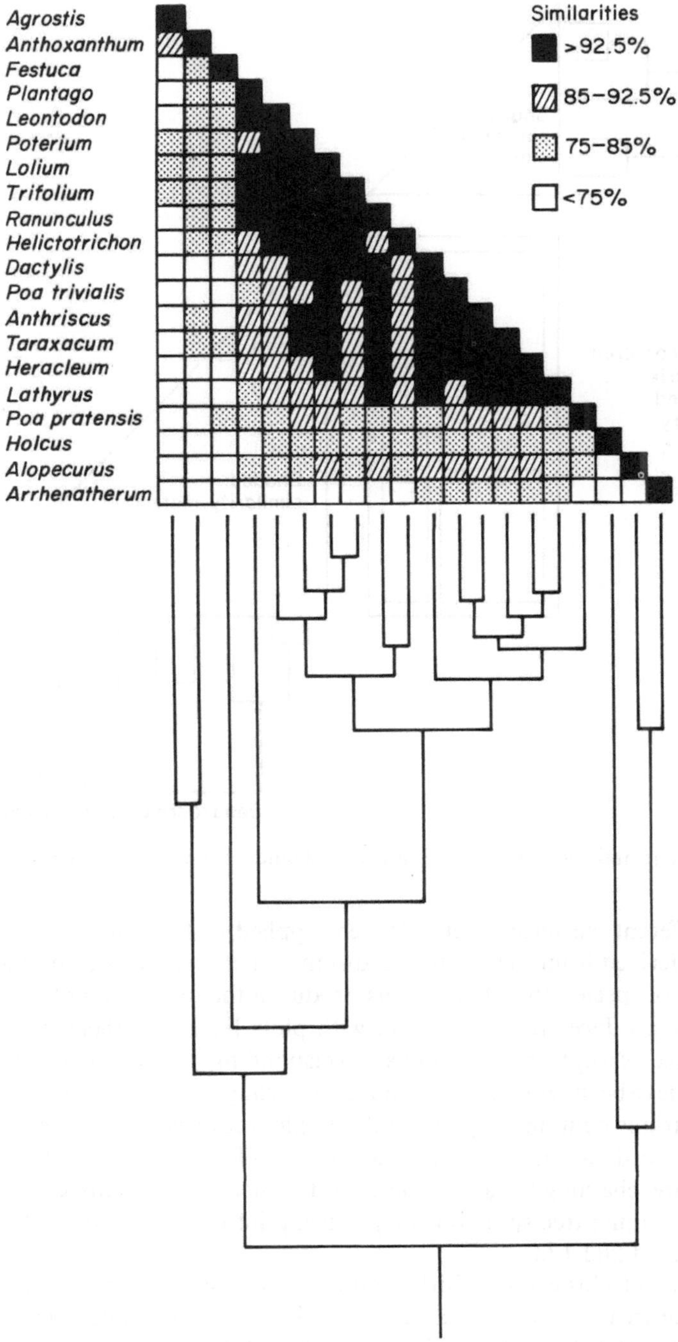

Fig. 5.7 Shaded similarity matrix for Park Grass species with hanging dendrogram (reordered from Fig. 5.5).

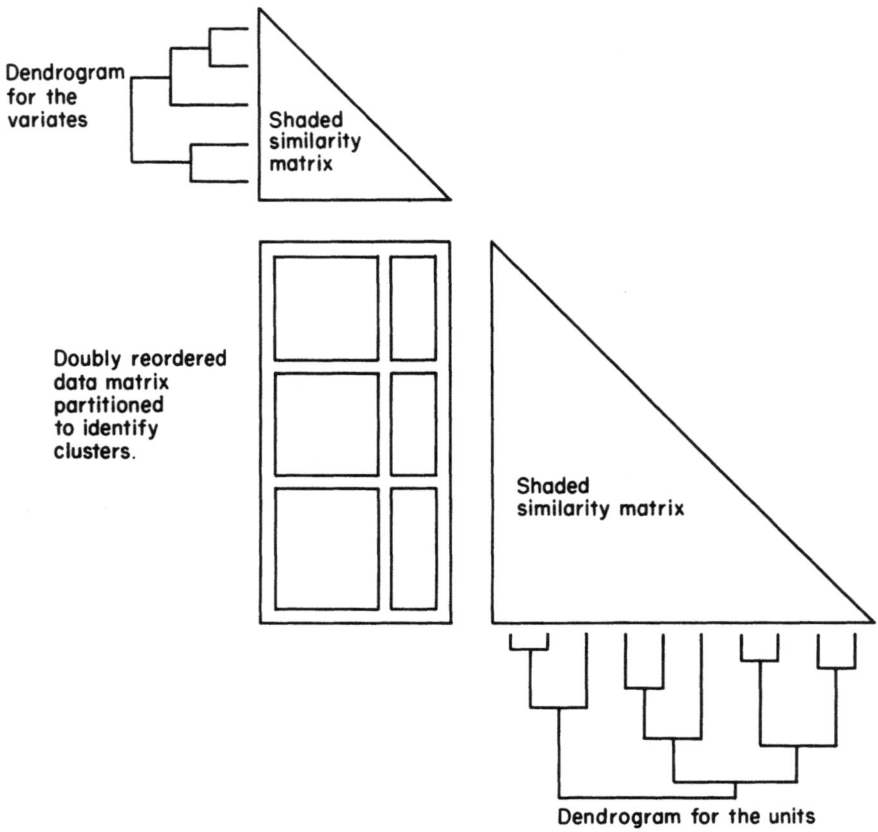

Fig. 5.8 A schematic way of combining row and column hierarchical analyses.

Three different clustering methods were applied to the similarity matrix, which was derived from the Euclidean distances between plots using the log percentages of species abundances: this produced the ordination of Fig. 3.6. Figure 5.9 reproduces this ordination with plots linked by their minimum spanning tree, except that the links corresponding to the three smallest similarities have been removed to produce four groups of plots of very unequal size. Two groups contain only one and two plots respectively, while a third straggles the ordination space and includes 21 of the 38 plots. This is an example of the chaining behaviour described earlier. The next division of the minimum spanning tree splits this major group into two by cutting the link joining plots 7d and 17d.

With centroid clustering a fairly natural classification into four groups occurs. Three are the same as three of the five single-linkage groups; the fourth combines the isolated plot 17a with the group of plots 3a, 3d, 8a, 8d and 17d.

Complete linkage clustering produces a very different classification of plots.

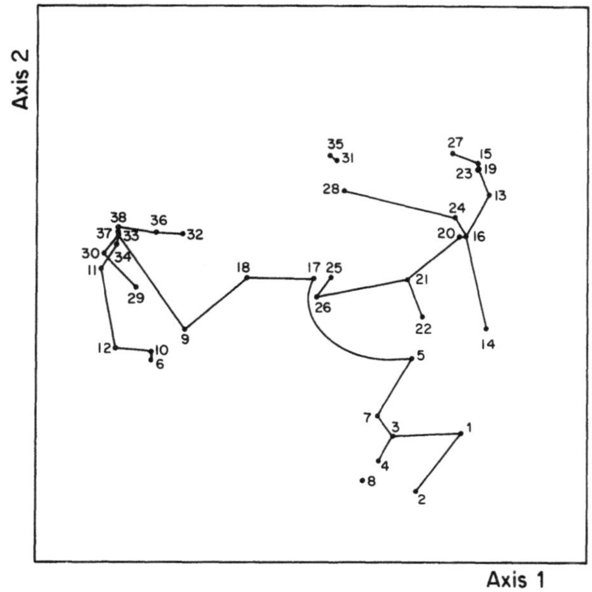

Fig. 5.9 Ordination of Park Grass plots (see Fig. 3.6) with segmented minimum spanning tree.

At 97% similarity, the method produces seven groups which appear reasonably compact in two-dimensional ordination space (Fig. 5.10). However, at a higher hierarchical level, group I appears to encircle group II. The grouping of plots 13c and 13d (which receive organic manure) and plot 7d (PK) with the unlimed, ammonium sulphate plots in group I is somewhat surprising. This association can be explained by looking at the furthest-neighbour similarities between plots, 13c, 13d and 7d, taken individually and as a group, and the groups Ia (plots 1d, 18d, 4d, 10d, 9d), Ib (9c, 11/1d, 11/2d), II, III, and IV.

Plot	Ia	Ib	II	III	IV
13c	96.4	96.2	95.9	94.5	94.9
13d	95.4	95.5	94.3	96.1	94.2
7d	96.0	94.8	96.7	93.3	96.4
Group similarity	95.4(13d)	94.8(7d)	94.3(13d)	93.3(7d)	94.2(13d)

Plots 13d and 7d, taken individually, would merge with groups III and II respectively, but taken together (and with plot 13c), the furthest neighbour of this group lies closest to the group Ia. The merging of this new group with group Ib occurs at the 94.8 similarity value given in the table above, since plot 7d is the furthest neighbour of the new group from group 1b. This is another

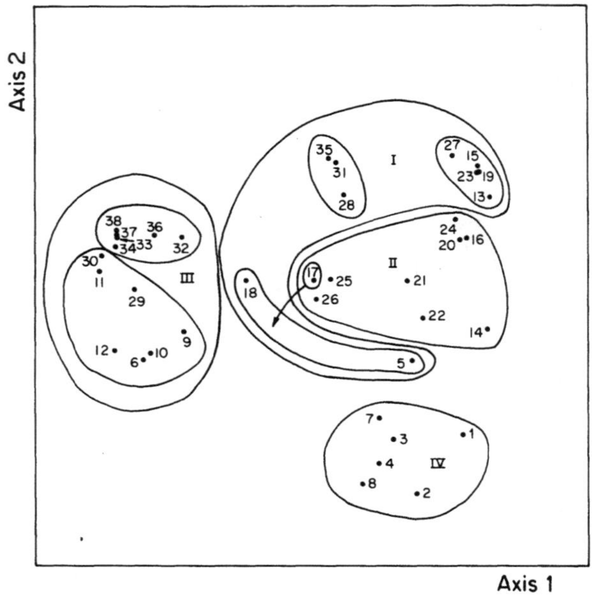

Fig. 5.10 **Ordination of Park Grass plots showing complete linkage clustering. Contours represent clusters at two different similarity thresholds.**

example of the shortcomings of agglomerative methods: an initial grouping has imposed undue constraint on later groupings.

To examine this further, the complete linkage classification into four groups was used as a starting point for a non-hierarchical classification based on minimizing the total within-group sum of squares (Section 5.3). As shown in Fig. 5.11, the grouping was similar to that from complete linkage clustering; however, the three plots under question in group I have been reallocated to groups II and III. This is more in accord with a natural division of ordination space. Group I now corresponds to unlimed plots receiving ammonium sulphate, while groups II and III contain lime-treated ammonium sulphate plots. Groups I, II and III are therefore associated with increasing soil pH while group IV contains plots of low productivity with a diverse flora. An indication of the typical species composition of these groups of plots may be obtained by drawing the group boundaries on the biplot given in Fig. 3.8.

The classification procedure was also repeated using the single-linkage clusters as a starting point. This produced a different optimum classification: groups III and IV were the same as in Fig. 5.11, but the combined group I + II was split diagonally rather than horizontally. However, the sum of squares criterion was larger than before, indicating that this solution was only at a local optimum: this emphasizes the importance of using more than one starting classification when using non-hierarchical methods.

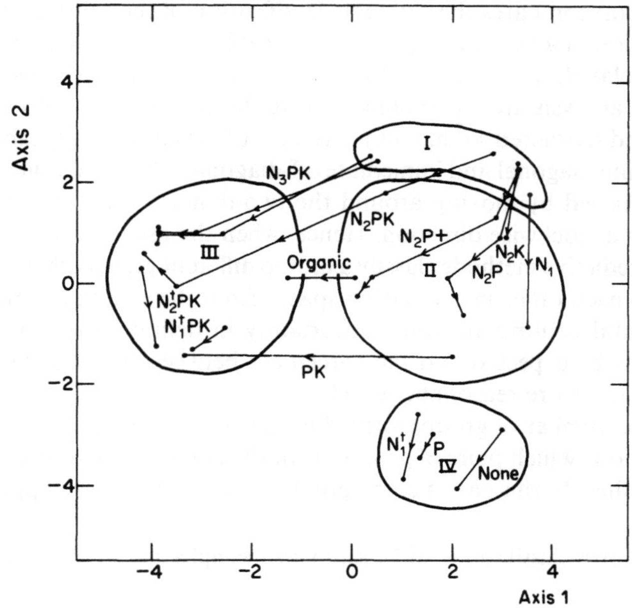

Fig. 5.11 Classification of Park Grass plots into four groups based on minimizing their total within-group sum of squares.

5.6 METHODS FOR COMPARING CLASSIFICATIONS

It is sometimes useful to compare a number of different classifications of the same set of units. For example, when faced with a plethora of possible methods for classifying samples, the field-worker may naturally wish to confirm that groupings produced by the different methods are in broad agreement: without some concordancy, one may question whether a natural grouping exists at all. Alternatively, an independent grouping of units based, for example, on environmental or morphometric information may be available for comparison.

The most straightforward way to compare two classifications is to form the two-way contingency table whose (i, j)th cell gives the number of individuals in group i from one classification and group j from the other. If the two classifications are in complete agreement there will be exactly one non-zero cell in each row and column of the table; in general these non-zero cells will not lie on the leading diagonal of the table, and it may then be convenient to rearrange the rows or columns so that they do. More generally, when two classifications do not completely match, it is easier to examine agreement and disagreement if the rows and columns are rearranged so that the larger entries in the table lie close to the leading diagonal.

Table 5.5(a) compares the original classification of the salt-marsh samples by association analysis into eight groups (Table 5.2) with the later maximal predictive classification (Table 5.4). In Table 5.5(b) the groups from the association analysis are recombined to give the classification into six groups (Fig. 5.3) and reordered so that non-zero cells of the contingency table lie only on the leading diagonal and immediate off-diagonals. Note that the new order may be achieved by moving around the two-dimensional ordination space (Fig. 5.4) in a clockwise direction. Hence, when the association analysis and maximal predictive methods classify a unit to different groups, these groups lie adjacent to each other in ordination space. Since the samples come from an environmental continuum, some uncertainty in defining the boundaries of groups must be expected and the agreement between the two classification methods is therefore reasonably good.

When the number of groups formed in each classification is large, it is not always obvious which pairs of groups from the two classifications are best to match together. In this case a correspondence analysis may be applied to the

Table 5.5 Cross-classification of 70 salt-marsh samples based on two methods of grouping.

(a) Original ordering of groups

		Classification by association analysis into 8 groups							
	Groups	I	II	III	IV	V	VI	VII	VIII
Maximal predictive classification	A	14	2	0	0	0	0	0	0
	B	0	0	10	0	0	0	0	0
	C	0	0	1	6	0	0	0	0
	D	0	0	0	0	3	0	0	0
	E	0	0	0	0	0	4	12	3
	F	0	8	0	0	0	1	0	6

(b) After combining and reordering groups

		Classification by association analysis into 6 groups					
	Groups	I+II	VIII	VI+VII	V	IV	III
Maximal predictive classification	A	16	0	0	0	0	0
	F	8	6	1	0	0	0
	E	0	3	16	0	0	0
	D	0	0	0	3	0	0
	C	0	0	0	0	6	1
	B	0	0	0	0	0	10

two-way table. Then rows and columns of the table that match should be represented by points close together on the plot; also a good ordering of the rows and columns is likely to be obvious from the ordination of the groups.

When more than two classifications are to be compared, inspection of the various two-way tables may well be adequate; if there are g separate groupings there will be $\frac{1}{2}g(g-1)$ such tables. Inspection of three- (and higher) way tables may also be useful, although interpretation is more difficult. Multiple correspondence analysis (Greenacre, 1984) may be helpful here.

Having rearranged the groups to give the closest associations between two classifications, a simple measure of their concordance is the proportion of units which are allocated to the same group in both classifications. Thus in Table 5.5(b), 57 of the 70 units (81%) are allocated to the same group in both classifications. When g classifications are being compared we may form a similarity matrix containing the proportions of matches for each of the $\frac{1}{2}g(g-1)$ pairs of classifications. A principal coordinates analysis may then be effective in displaying the similarities between the methods on a two-dimensional plot (cf. Section 4.3); or the classifications may themselves be classified from these similarities!

6 Analysis of asymmetry

In some studies of association between pairs of objects the measure of association is not symmetric. For example, in Table 6.1, the values x_{ij} measure the spatial association between pairs of tree species in Lansing Wood, Michigan, in terms of the number of times a tree of species j was the nearest neighbour of species i. As we see, x_{ij} need not equal x_{ji}, and hence the principal coordinates method of analysis of Chapter 3 is not immediately usable, as it requires the association matrix to be symmetric. One solution is to construct a symmetric matrix of associations for species i and j by taking the mean of x_{ij} and x_{ji}: this would involve little loss of information from Table 6.1, where asymmetry appears to be due to the chance effects of sampling (see Section 6.1). Consider, however, the data of Table 6.2 which shows the pattern of pecking behaviour among a flock of 12 chickens. Here the information on social hierarchy is wholly contained in the asymmetry and would be lost if values of x_{ij} and x_{ji} were simply averaged. In another example considered in Section 6.3, an association matrix records the number of quadrats in which one vegetation type succeeds another in time; the asymmetry is then found to give important information about the ecological succession.

Table 6.1 Spatial association among tree species in Lansing Wood measured by the number of times each species occurs as nearest neighbour.

	Number of occurrences as nearest neighbour						
Species	Red oak	White oak	Black oak	Hickory	Maple	Other	Total trees
Red oak	104	59	14	95	64	10	346
White oak	62	138	20	117	95	16	448
Black oak	12	20	27	51	25	0	135
Hickory	105	108	48	355	71	16	703
Maple	74	70	21	79	242	28	514
Other	11	14	0	25	30	25	105

In this chapter we consider methods for analysing asymmetric matrices. The first approach uses the established techniques developed previously for rectangular (species-by-sites) matrices. The second approach, developed by Gower (1977), partitions the matrix into symmetric and skew-symmetric parts and involves a novel planar representation of the latter.

Table 6.2 Pecking relationships in a flock of 12 hens showing a perfect linear hierarchy. The elements x_{ij} of the matrix indicate the number of times the hen in the ith row was observed to peck the hen in the jth column. (From Guhl, 1956.)

		Subordinate hen											
		Y	B	V	R	G	YY	BB	VV	RR	GG	YB	BR
	Y	0	22	8	18	11	30	10	12	17	6	11	21
	B	0	0	29	11	21	7	12	17	26	16	7	6
	V	0	0	0	6	11	6	3	27	8	7	2	16
Dominant	R	0	0	0	0	12	21	8	6	6	26	17	3
hen	G	0	0	0	0	0	8	15	3	10	8	12	15
	YY	0	0	0	0	0	0	30	19	17	6	13	8
	BB	0	0	0	0	0	0	0	8	3	12	11	12
	VV	0	0	0	0	0	0	0	0	13	26	18	20
	RR	0	0	0	0	0	0	0	0	0	6	8	12
	GG	0	0	0	0	0	0	0	0	0	0	21	6
	YB	0	0	0	0	0	0	0	0	0	0	0	27
	BR	0	0	0	0	0	0	0	0	0	0	0	0

6.1 ROW AND COLUMN PLOTS

This entails plotting the objects twice on the same graph, once as the rows and once as the columns of the association matrix. The biplot technique (Section 3.2.3) provides one way of doing this. The matrices of coordinates **A** and **B** are derived from the singular value decomposition of the association matrix, $\mathbf{X}=\mathbf{USV}'$; setting $\mathbf{A}=\mathbf{US}^{1/2}$ and $\mathbf{B}=\mathbf{VS}^{1/2}$, which places them on the same scale. Now when **X** is symmetric, and positive semi-definite (p.s.d.), $\mathbf{U}=\mathbf{V}$, and hence the coordinates for rows and columns are identical. The degree of asymmetry is therefore indicated by the lack of coincidence among the pairs of points. (When **X** is symmetric, but not p.s.d., **A** and **B** will differ, but only by a change of sign in one or more columns; however, unless **X** is degenerate, these columns will contain little of the variation in **X**. In practice, for the first few dimensions of an ordination, **A** and **B** will be identical.)

Figure 6.1 shows the two-dimensional biplot for the Lansing Wood association matrix. The interpretation follows from applying the inner product rule as before. Thus the number of times white oak (W) is the nearest neighbour of hickory (H) is approximated by the inner product of vectors OH and Ow, while the reciprocal association is given by the inner product of OW and Oh. Here upper and lower case letters represent points for species (closed circles) and neighbours (open circles), respectively, in Fig. 6.1. The points of each pair are seen to be almost coincident, indicating little asymmetry, as is evident from Table 6.1: Pielou (1961) suggests lack of symmetry in the nearest-neighbour relationship may occur if the species differ markedly in size or in the degree of local aggregation. Even when the asymmetry is ignored, an analysis

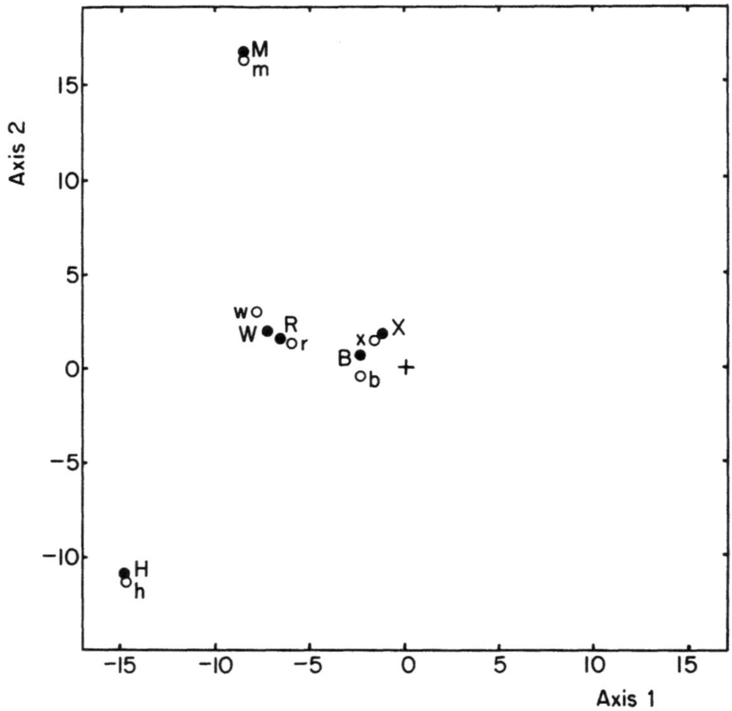

Fig. 6.1 Biplot of tree species and their nearest neighbours in Lansing Wood, for the association matrix in Table 6.1. The six species are each plotted as a pair of points representing their position as a species (solid circles) and as a nearest neighbour (open circles). Species codes are as follows; R(r) red oak; W(w) white oak; B(b) black oak; H(h) hickory; M(m) maple; X(x) other species. + indicates the origin.

of the associations between species, using the diagnostic procedures of Section 3.2.5, is still revealing. Under the assumption that all species are randomly and independently distributed, the expected number of times species j occurs as nearest neighbour to species i is Np_ip_j; where N is the total number of trees sampled, and p_i and p_j are the frequencies of species i and j. The two-dimensional biplot is then exact and the species points P_i $(i=1, \ldots, n)$ lie on a line passing through the origin O such that $|OP_i| = \sqrt{Np_i}$ $(i=1, \ldots, n)$. (This is a special case of the multiplicative model displayed in Fig. 3.9(b).) Hence the degree of segregation between species i and j is indicated by the angle subtended by the points P_i, P_j at the origin. Consider, for example, the association between the two most abundant species in the wood, hickory (H) and maple (M). Figure 6.1 shows the vectors OH (Oh) and Om (OM) to have the largest magnitudes, as expected, but to be almost at right angles to each other so that their inner product is relatively small. This indicates a high

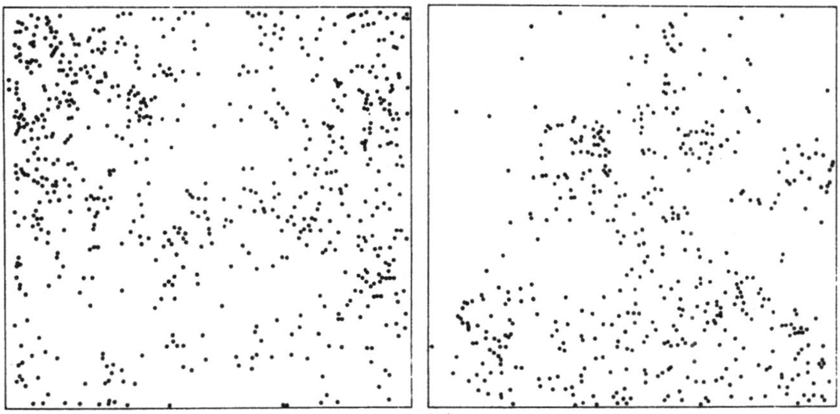

(a) (b)

Fig. 6.2 Distribution of (a) hickories and (b) maples in an 8 hectare plot from Lansing Wood, Michigan. (Reproduced with permission from Fig. 17 of Diggle, 1983.)

degree of segregation between the two species, which is immediately apparent from their contrasting spatial distributions shown in Fig. 6.2.

Correspondence analysis (Section 3.3) is an alternative method of plotting rows and columns of an association matrix on the same graph. It is used here to refine a preliminary ordination of prominent tree species from 104 forest stands in Wisconsin (Brown and Curtis, 1952). The species associations of Table 6.3 were obtained by first classifying the stands according to the most

Table 6.3 Average importance value of trees in stands with given species as leading dominant. (From Brown and Curtis, 1952.)

		Species								
Number of stands	*Leading dominant*	*As*	*Tc*	*Qr*	*Bp*	*Ps*	*Pr*	*Pt*	*Qe*	*Pb*
23	*Acer saccharum*	145	25	22	6	1	0	1	0	0
23	*Tsuga canadensis*	40	152	3	5	4	3	0	0	0
6	*Quercus rubra*	27	1	138	23	10	8	5	3	0
6	*Betula papyrifera*	48	8	16	108	19	1	29	1	0
19	*Pinus strobus*	12	6	12	12	150	39	9	5	0
9	*Pinus resinosa*	3	0	15	14	56	156	24	4	2
4	*Populus tremuloides*	11	0	29	34	14	19	140	0	0
4	*Quercus ellipsoidalis*	0	0	7	1	11	9	9	103	56
10	*Pinus banksiana*	0	0	3	3	13	12	14	36	213
Average importance value (104 stands)		48	41	20	15	37	24	13	9	23

dominant species, and then calculating the average importance of each species in each class. Figure 6.3 shows the two-dimensional plot for species rows and columns from a correspondence analysis. Once again the pairs of points for each species appear close together; this indicates that the asymmetry, although more pronounced than in the previous example, does not distort the main ordination representing a transition from pioneer to climax species. This species sequence is very similar to that derived somewhat subjectively by Brown and Curtis (1952); except that they chose *Acer saccharum* as the pre-eminent climax species while the correspondence analysis suggests that this role is taken by *Tsuga canadensis*.

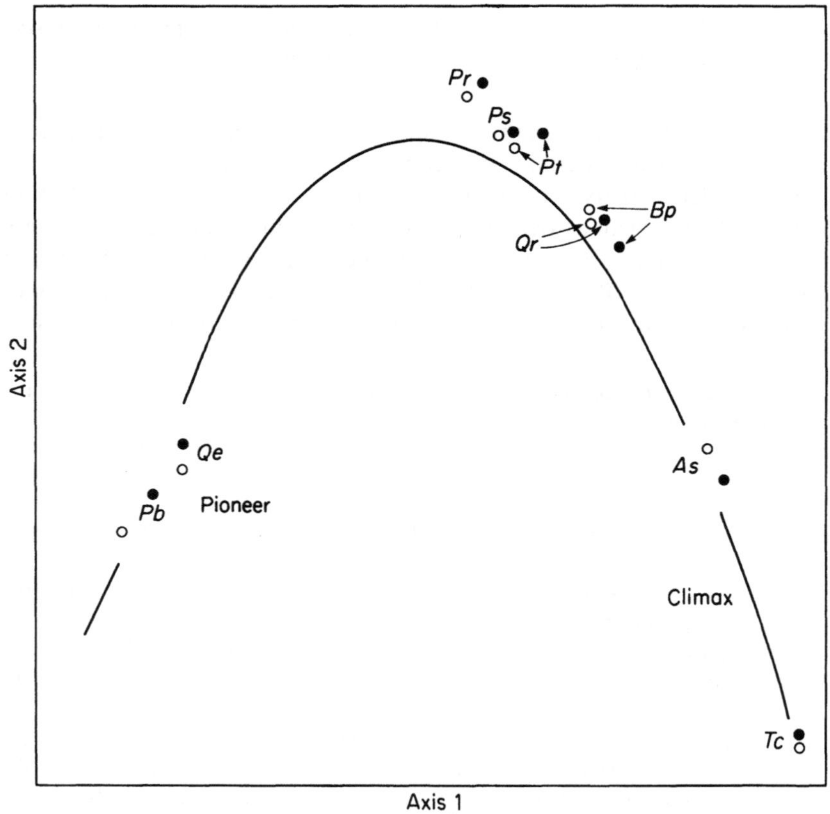

Fig. 6.3 Correspondence analysis of associations among tree species from forest stands in Wisconsin, showing a gradient from pioneer to climax species. Each species is plotted as a pair of closed and open circles representing respectively the rows and columns of Table 6.3.

6.2 SKEW-SYMMETRY ANALYSIS

An overall analysis of the association matrix **X**, like those of the previous section, confounds the symmetric and asymmetric components of the data and this will often obscure a more detailed investigation of the asymmetry. Gower (1977) therefore proposed partitioning the associations x_{ij}, before analysis, into an average association, $d_{ij}=\frac{1}{2}(x_{ij}+x_{ji})$, and the difference from that average, $z_{ij}=x_{ij}-d_{ij}=\frac{1}{2}(x_{ij}-x_{ji})$. Thus

$$\mathbf{X}=\mathbf{D}+\mathbf{Z}. \tag{6.1}$$

Clearly **D** is symmetric and can be analysed by the methods of Chapter 3, while **Z** is skew-symmetric, i.e. $z_{ij}=-z_{ji}$. This partitioning of the data matrix is particularly appropriate when different mechanisms give rise to the symmetric and skew-symmetric parts, as is often found to be the case.

To analyse **Z** we again use the singular-value decomposition, $\mathbf{Z}=\mathbf{USV}'$. A biplot interpretation is again possible but, for the special case when **Z** is skew-symmetric, it can be shown that $\mathbf{V}=\mathbf{JU}$ where **J** is a block-diagonal skew-symmetric matrix made of of blocks $\begin{pmatrix} 0 & 1 \\ -1 & 0 \end{pmatrix}$: this allows a more succinct description.

The elements of **U** and **V** are now related in pairwise fashion, $v_{i1}=u_{i2}$, $v_{i2}=-u_{i1}$, $v_{i3}=u_{i4}$, $v_{i4}=-u_{i3}$, etc.; the singular values also come out in pairs, $s_1=s_2$, $s_3=s_4$, and so on, where, if n is an odd number, $s_n=0$. The elements of **Z** may thus be expanded as a sum of skew-symmetric components

$$z_{ij}=s_1(u_{i1}u_{j2}-u_{i2}u_{j1})+s_3(u_{i3}u_{j4}-u_{i4}u_{j3})+\ldots. \tag{6.2}$$

The best two-dimensional fit is

$$\hat{z}_{ij}=s_1(u_{i1}u_{j2}-u_{i2}u_{j1}), \tag{6.3}$$

which accounts for a proportion $2s_1^2/\Sigma_1^n s_k^2$ of the total variation in **Z**. This may be an adequate representation of **Z**; otherwise extra pairs of dimensions (planes) may be included.

As in previous analyses, each planar fit can be plotted: for example, the first two dimensions are represented by a graph with origin O containing the points P_i at (u_{i1}, u_{i2}) for $i=1, 2, \ldots, n$. However, for this analysis, points P_i and P_j are compared, not by the distance between them, but by the area of the triangle OP_iP_j which is proportional to the right-hand side of equation (6.3) (for proof see Section 6.4). Generally we shall only be concerned with estimating the relative magnitudes of the skew-symmetric elements z_{ij}: to derive their absolute values directly from the two-dimensional plot we must use the scaled coordinates (a_{i1}, a_{i2}), where $a_{i1}=\sqrt{(2s_1)}u_{i1}$ and $a_{i2}=\sqrt{(2s_1)}u_{i2}$. Note also that $z_{ij}=z_{ji}$ and thus the areas of the triangles OP_iP_j and OP_jP_i are considered equal, but opposite in sign (see below).

This triangle-area interpretation is novel and not such an easy concept as the distance interpretation used previously, so we offer some guidelines to its use. (Remember that the area of a triangle is equal to half its base times its perpendicular height.) Table 6.4 gives the singular value decomposition for a hypothetical association matrix for four species. The skew-symmetric elements z_{ij} have been especially chosen so that the two-dimensional representation given in Fig. 6.4 is almost exact. The graph is interpreted as follows: since P_1 and P_3 are close one may assume that z_{13} (=16) is small; however, P_1 and P_2 are distant but roughly collinear with the origin O, so z_{12} (=10) is also small; in contrast, P_1 and P_4 are distant but subtend almost a right angle at the origin, so z_{14} (=60) is large; finally, the sign of z_{ij} is indicated

Table 6.4 Hypothetical association matrix for four species, P_1, P_2, P_3, P_4, and derived skew-symmetric matrix with singular value decomposition.

Association matrix, **X**

	P_1	P_2	P_3	P_4
P_1	32	42	35	141
P_2	22	102	44	82
P_3	3	56	51	140
P_4	21	62	38	129

Skew-symmetric part, **Z**

	P_1	P_2	P_3	P_4
P_1	0	+10	+16	+60
P_2	−10	0	−6	+10
P_3	−16	+6	0	+51
P_4	−60	−10	−51	0

The singular value decomposition, $\mathbf{Z}=\mathbf{USV}'$ gives

$$\mathbf{S}=\begin{bmatrix} 81.8 & 0 & 0 & 0 \\ 0 & 81.8 & 0 & 0 \\ 0 & 0 & 0.1 & 0 \\ 0 & 0 & 0 & 0.1 \end{bmatrix}, \quad \mathbf{U}=\begin{bmatrix} 0.00 & -0.77 & 0.64 & 0.00 \\ 0.16 & -0.10 & -0.12 & -0.98 \\ 0.25 & -0.61 & -0.73 & 0.19 \\ 0.95 & 0.18 & 0.21 & 0.11 \end{bmatrix},$$

$$\mathbf{V}=\begin{bmatrix} -0.77 & 0.00 & 0.00 & -0.64 \\ -0.10 & -0.16 & -0.98 & 0.12 \\ -0.61 & -0.25 & 0.19 & 0.73 \\ 0.18 & -0.95 & 0.11 & -0.21 \end{bmatrix}$$

The coordinates of P_1, P_2, P_3 and P_4 in the two-dimensional representation (Fig. 6.4) are given by the first two columns of **U** multiplied by $\sqrt{2s_1}=12.8$; viz $P_1=(0, -9.9)$, $P_2=(2.0, -1.3)$, $P_3=(3.2, -7.8)$, $P_4=(12.2, 2.3)$.

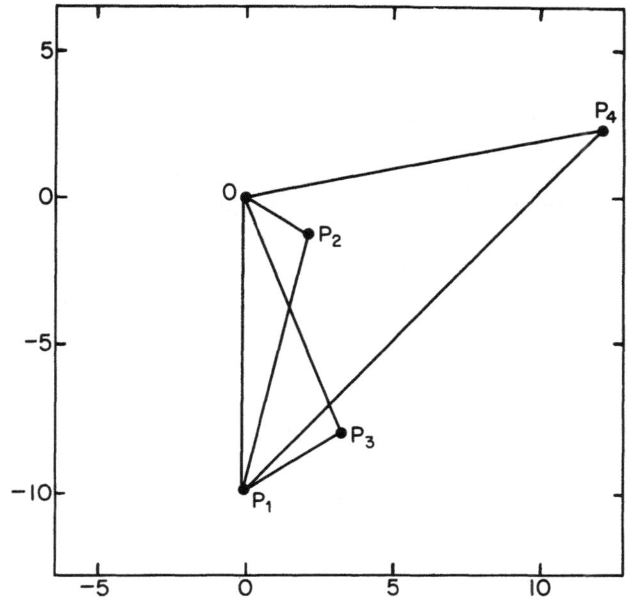

Fig. 6.4 Triangle-area representation of the skew-symmetric part of the transition matrix in Table 6.4.

by the direction of circumscribing the triangle OP_iP_j, negative for a clockwise direction, positive for anticlockwise. This is consistent in any one diagram; however, inspection of the data will be necessary to establish which direction corresponds to which sign in each plane.

In general, the two-dimensional representation will not be exact, and additional information about the skew symmetry may be gained from plotting the second pair of dimensions (3 and 4), or further pairs of dimensions. (The dimensions are only interpretable in pairs, so it is not fruitful to plot, for example, u_{i1} against u_{i3}.) An indication of the adequacy of the two-dimensional approximation is obtained from the squared singular values in the usual way.

We now illustrate the graphical representation for two simple forms of skew-symmetric matrix.

6.2.1 Analysis of matrices for social hierarchies

The hens classifying the matrix of peck-dominance in Table 6.2 obey a simple linear hierarchy. More commonly, flocks will show some deviation from linearity but it is often still useful to derive a rank order of individuals: here an analysis of the skew-symmetric component of the dominance matrix may prove helpful. We shall consider this problem as a case study in the next section, but first describe the graphical representation obtained for two dominance matrices showing a perfect linear hierarchy.

Table 6.5(a) shows the simple dominance matrix derived from Table 6.2 in which $x_{ij}=1$ if hen i dominates j, and 0 otherwise. This produces a skew-symmetric matrix with $+\frac{1}{2}$ above and $-\frac{1}{2}$ below the diagonal and a graphical representation in the first plane which preserves the ordering of individuals, although the points lie on an arc rather than a straight line (Fig. 6.5(a)). Following Gower (1977), it can be shown that the scaled points (a_{i1}, a_{i2}) lie on a circle about the origin with radius $2\sqrt{(s_1/n)}$ and that successive points subtend an angle of π/n at the centre, where n is the number of individuals. The

Table 6.5 Two artificial dominance matrices showing a perfect linear hierarchy.

(a)

Dominant \ *Subordinate*	1	2	3	4	5	6	7	8	9	10	11	12
1	0	1	1	1	1	1	1	1	1	1	1	1
2	0	0	1	1	1	1	1	1	1	1	1	1
3	0	0	0	1	1	1	1	1	1	1	1	1
4	0	0	0	0	1	1	1	1	1	1	1	1
5	0	0	0	0	0	1	1	1	1	1	1	1
6	0	0	0	0	0	0	1	1	1	1	1	1
7	0	0	0	0	0	0	0	1	1	1	1	1
8	0	0	0	0	0	0	0	0	1	1	1	1
9	0	0	0	0	0	0	0	0	0	1	1	1
10	0	0	0	0	0	0	0	0	0	0	1	1
11	0	0	0	0	0	0	0	0	0	0	0	1
12	0	0	0	0	0	0	0	0	0	0	0	0

(b)

Dominant \ *Subordinate*	1	2	3	4	5	6	7	8	9	10	11	12
1	0	1	2	3	4	5	6	7	8	9	10	11
2	0	0	1	2	3	4	5	6	7	8	9	10
3	0	0	0	1	2	3	4	5	6	7	8	9
4	0	0	0	0	1	2	3	4	5	6	7	8
5	0	0	0	0	0	1	2	3	4	5	6	7
6	0	0	0	0	0	0	1	2	3	4	5	6
7	0	0	0	0	0	0	0	1	2	3	4	5
8	0	0	0	0	0	0	0	0	1	2	3	4
9	0	0	0	0	0	0	0	0	0	1	2	3
10	0	0	0	0	0	0	0	0	0	0	1	2
11	0	0	0	0	0	0	0	0	0	0	0	1
12	0	0	0	0	0	0	0	0	0	0	0	0

singular values of successive planar representations are given by $s_k = \frac{1}{2}\tan\{(n-k)\pi/2n\}$, for $k=1, 3, 5, 7, \ldots, (n-1)$, so for $n=12$, 87.4% of the variation is described by the first planar representation shown in Fig. 6.5(a).

An alternative form of dominance matrix is given by Table 6.5(b) where $x_{ij}=c(j-i)$ for $j>i$, and 0 otherwise (here $c=1$). The two-dimensional representation (Fig. 6.5(b)) is now exact and provides equally spaced points on a straight line, L. This result is easily explained using the triangle-area rule: the height from the origin for any triangle OP_iP_j, whose base P_iP_j lies along L, is a constant and hence the area of the triangle OP_iP_j is proportional to the distance P_iP_j of the points apart, i.e. to $|i-j|$.

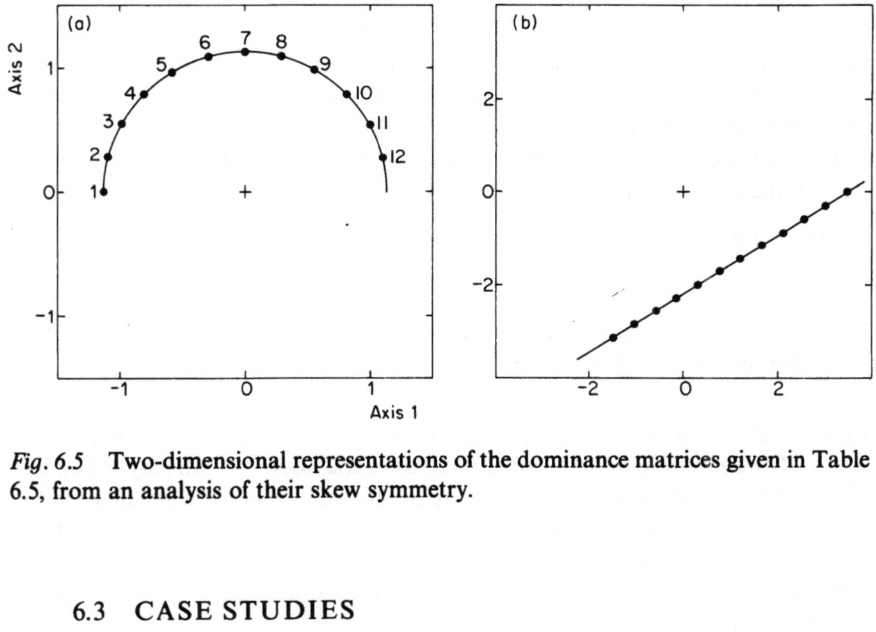

Fig. 6.5 Two-dimensional representations of the dominance matrices given in Table 6.5, from an analysis of their skew symmetry.

6.3 CASE STUDIES

We now describe three different examples of analysing asymmetric data. In the first, a skew-symmetry analysis is used to identify an approximate pecking order among a flock of birds and to describe the changes in order that are observed with time. An analysis of skew symmetry is also used in the second example to illustrate the major successional changes which take place in heathland vegetation during regeneration after fire. The final example describes the competitive interactions of nine grassland species when grown together in pairs. Here a biplot display of the matrix of log yields suggests a simple model for describing species interactions which is subsequently fitted to the data. Further examples of the possible application of skew-symmetry analysis in ecology are given by Mitchley and Guarino (1984) and Constantine and Gower (1982).

6.3.1 Analysis of pecking relationships among domestic fowl

Guhl (1953) gives the pecking relationships among a flock of 32 hens. In Table 2.1(b) the birds were ordered according to the number of birds over which each is dominant (the pecking score) and they are seen to depart from a simple linear hierarchy. An example of the complex pecking relationships which exist is shown below for birds 2, 3, 4 and 5.

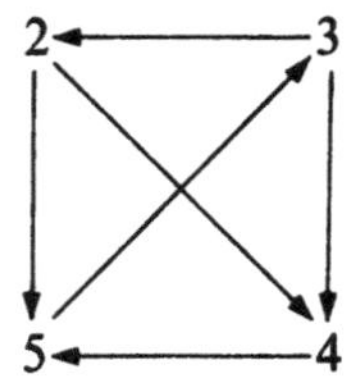

Whatever ordering is chosen for these four birds, there are at least two reversions in which one bird pecks another higher in the order. This is apparent from the two triangular relationships 2, 5, 3 and 3, 4, 5.

In the absence of a perfect linear hierarchy, a suitable ordering of birds could be obtained by minimizing:

(i) the total number of reversions, $\sum_{i>j} x_{ij}$;

or

(ii) the number of reversions weighted by the separation of the relevant birds in the hierarchy, $\sum_{i>j} (i-j)x_{ij}$.

In Table 2.1 we used criterion (i) to order birds which have the same pecking score. However, unless the number of birds in the flock is small (say <10), it is not feasible to minimize (i) or (ii), for all birds, by looking at all possible orderings. A computer algorithm is then required to search for the solution: this involves the same problem of identifying a global minimum among possibly many local minima as occurs with maximal predictive classification (Section 5.3).

As an alternative approach, we assess here the performance of an analysis of the skew symmetry of the matrix of Table 2.1. The representation of points (a_{i1}, a_{i2}) in the first plane is shown in Fig. 6.6, together with the fitted arc expected for a simple linear hierarchy. The ordering of birds may be better visualized using a transformed plot (Fig. 6.7) in which the distance of the points from the origin in Fig. 6.6 is plotted against the angle made with the horizontal. (Note, however, that the triangle-area interpretation does not apply to this graph.) Individuals now showing large deviations from the vertical line are those that show major reversions in pecking order. Figure 6.8 shows that the new ordering of hens is similar to that based on pecking score (Table 2.1): the total number of reversions has been reduced from 42 to 36 but at the cost of increasing the value of the weighted criterion (ii).

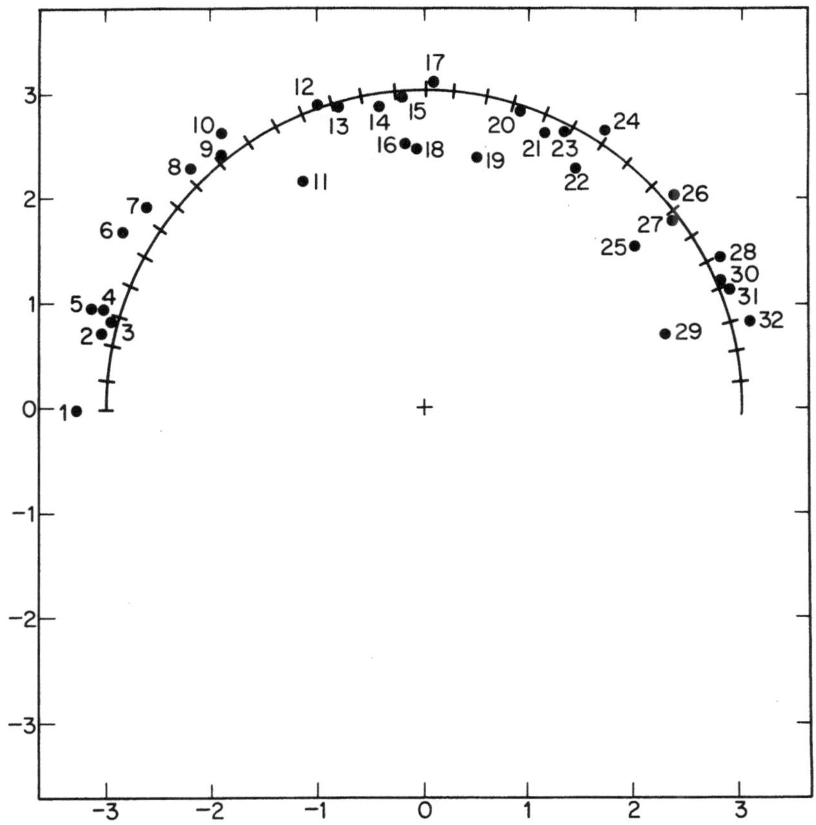

Fig. 6.6 Two-dimensional representation of pecking relationships among a flock of 32 hens (Table 2.1). An approximate linear order may be derived by projecting the points onto the semicircle shown.

Figure 6.8 also shows the ordering obtained using a simple switching algorithm defined as follows: starting with some initial ordering and the lowest ranking bird, move up the table interchanging the order of adjacent pairs of birds when the lower ranking bird is dominant – i.e. interchange birds i and $i-1$ if $x_{i,i-1}=1$; repeat this procedure until no more exchanges are possible. In general the final ordering of birds will depend on the chosen initial ordering, but here the same result was achieved starting with the ordering based on either the pecking score or the skew-symmetry analysis, although the latter took fewer iterations. Using this algorithm, the number of reversions is reduced to 29, which we suspect is close to the minimum. However, the ordering from the skew-symmetry analysis produced a much smaller weighted reversion score and may be preferred in some circumstances.

In many studies of groups of animals the dominance relationship may be

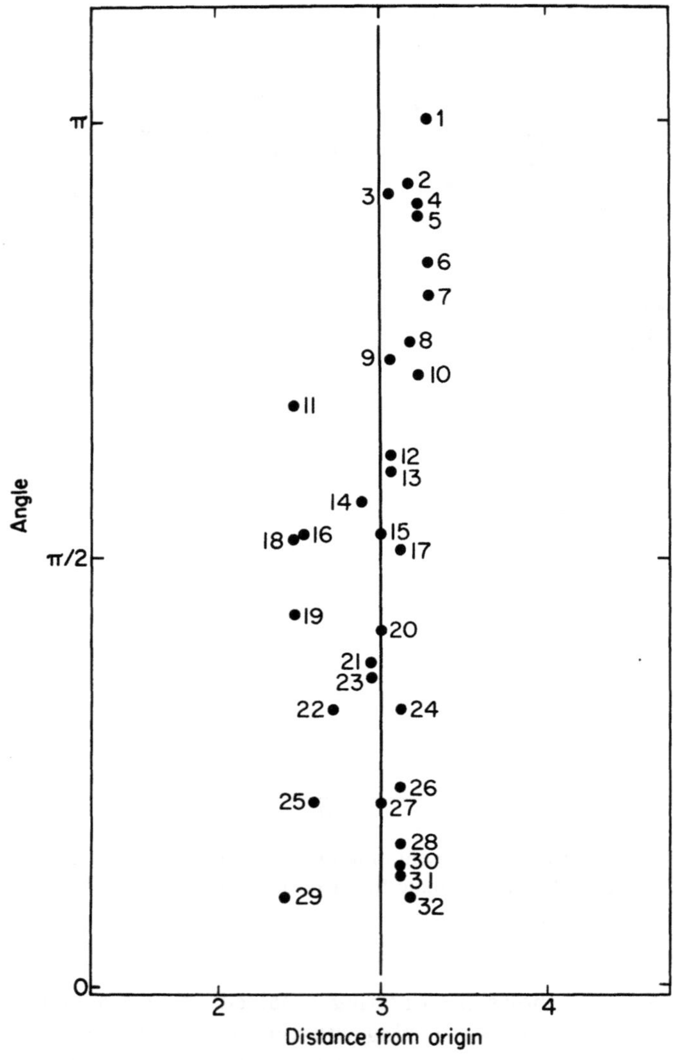

Fig. 6.7 Transformed plot of Fig. 6.6 using polar coordinates, to show an approximate linear hierarchy among the 32 hens.

ambiguous or unknown. In Table 6.2 we scored those paired relationships where pecking appeared to be in both directions as ($\frac{1}{2}$, $\frac{1}{2}$). (Note that Guhl (1953) makes no mention of such behaviour and there could be minor transcription errors in his original table.) Other workers have also scored unknown dominance relationships as ($\frac{1}{2}$, $\frac{1}{2}$). This is rarely correct, since in many cases the dominance relationship between two individuals will be unrecorded because of the deliberate avoidance of contact by the subordinate

Fig. 6.8 Ranking of flock of 32 hens based on three different procedures (see text).

	Pecking score	*Skew-symmetry analysis*	*Switching adjacent pairs*
	1	1	1
	2	2	3
	3	3	2
	4	4	4
	5	5	5
	6	6	6
	7	7	7
	8	8	8
	9	9	9
	10	10	10
	11	11	12
	12	12	11
	13	13	16
	14	14	13
	15	16	14
	16	15	15
	17	18	18
	18	17	17
	19	19	19
	20	20	20
	21	21	21
	22	22	22
	23	23	23
	24	24	24
	25	27	27
	26	25	25
	27	26	28
	28	28	26
	29	30	30
	30	31	31
	31	29	32
	32	32	29
No. of reversions	42	36	29
Weighted reversion score	163	$168\frac{1}{2}$	$181\frac{1}{2}$

individual towards the more dominant. The correct treatment of missing values in skew-symmetry analysis would require special computer algorithms for weighted least squares which unfortunately are not yet widely available in multivariate computer packages (see Chapter 7).

Ambiguous dominance relationships, in which individual *i* may sometimes beat individual j, while at other times j beats i, may be subjected to analysis of their skew symmetry by identifying x_{ij} with the proportion of encounters in which i dominates j. Ambiguous dominance relationships can arise from lumping together observations from different sites in a habitat, where the dominance relationship is clear cut at each site but varies between sites (Brown, 1975). In these situations a satisfactory ordering of individuals in one dimension is unlikely. As an example we consider a study by Masure and Allee (1934) of pecking relationship in two flocks of seven male and seven female pigeons. The two flocks were observed on several occasions before being brought together to mate. After mating, the flocks were separated again and further observations made on their pecking behaviours. The results for the separate observation dates were remarkably consistent (save for the fourth date for females; see later) but the pecking relationships among both males and females changed after the mating period (Table 6.6). These changes are demonstrated by applying a skew-symmetry analysis to the pecking

Table 6.6 Pecking frequencies among seven male and seven female pigeons and after mating. Table entries show the number of times that the pigeon specified by row pecks that specified by column. (From Masure and Allee, 1934, summing over all observation dates before and after sampling, except the fourth date for females.)

FEMALES

	Before mating							*After mating*						
	BB	BR	BW	BY	GW	RW	RY	BB	BR	BW	BY	GW	RW	RY
BB	0	41	44	68	58	29	45	0	24	77	17	115	33	60
BR	6	0	18	40	16	3	11	1	0	3	0	63	0	8
BW	10	7	0	11	5	18	12	33	9	0	8	18	9	5
BY	62	44	62	0	66	49	31	27	68	50	0	87	99	47
GW	36	20	23	28	0	29	11	26	9	35	18	0	27	31
RW	0	9	8	22	10	0	12	3	4	5	10	8	0	0
RY	10	2	5	2	5	5	0	143	109	142	227	145	136	0

MALES

	Before mating							*After mating*						
	B	BL	G	R	W	Y	YY	B	BL	G	R	W	Y	YY
B	0	0	10	4	4	26	0	0	0	0	2	0	0	0
BL	35	0	39	53	39	33	73	18	0	15	57	22	6	14
G	38	43	0	55	33	42	36	14	37	0	58	79	25	44
R	57	31	40	0	48	49	25	10	16	29	0	23	16	13
W	4	7	13	22	0	11	61	2	7	2	4	0	0	6
Y	32	19	35	58	28	0	55	14	48	34	41	38	0	37
YY	60	29	88	93	63	31	0	20	49	92	37	49	18	0

frequencies, after standardizing for the number of pair contacts; i.e. the elements of Table 6.6 are first transformed so that $y_{ij}=x_{ij}/(x_{ij}+x_{ji})$. Figure 6.9 shows the separate two-dimensional plots of bird coordinates, before and after mating, interposed on the same graph. The more dominant birds appear at the top of the graph and arrows indicate the more subordinate bird of a pair based on their pecking relationship; where no arrow is shown, the dominance relationship is that expected from the relative positions of the two birds.

For female birds the two-dimensional representations (Fig. 6.9(a)) account for 80% and 83% of the variation for the pre-mated and post-mated flock respectively. Comparison of the plots show the dramatic rise in dominance of bird RY: this bird, together with BY, was not mated when the flocks were brought together, although it had already begun to dominate birds GW, RW and BR by the fourth observation time which was immediately prior to mating. (Observations for this time were consequently omitted from the cumulative frequencies of Table 6.6.) The four subordinate birds, BR, BW, GW and RW, showed much less pecking contact among themselves than other birds and their social ordering is poorly defined, especially after mating.

The ordering of the male birds is also poorly defined before mating, but afterwards the males form a linear hierarchy, well represented in the right-hand side of Fig. 6.9(b) which accounts for 95% of total variation; here the two unmated birds, Y and B, appear at either end of the social order.

6.3.2 Plant species succession

Hobbs and Legg (1984) describe the changes in vegetation of heathland during regeneration after burning. In one study, random 10 cm × 10 cm quadrats were scored for evidence of change, past or present, from one vegetation state to another and a transition matrix (Table 6.7) was constructed. Following Usher (1981), we can draw a diagram showing the main transitions between states (Fig. 6.10) which suggests a cyclical climax involving four species:

Erica→*Calluna*→*Hypnum*→*Arctostaphylos*→*Erica* or *Calluna*.

We now see how closely this pattern is displayed by a formal analysis of the skew symmetry of the transition matrix. The states in Table 6.7 have been especially arranged to emphasize the asymmetry of the matrix. It is notable that, save for the last column, most of the elements above the diagonal are zeros, i.e. transitions between pairs of states are for the most part unidirectional. Now when the matrix is partitioned as in equation (6.1), the respective elements of the symmetric and skew-symmetric matrices are approximately equal in size (save for diagonal elements) but the symmetric component excludes crucial information on the direction of the transitions. Hence we shall be concerned here only with analysing the skew-symmetric component of the transition matrix.

A skew-symmetry analysis gave singular values, 35, 35, 16, 16, 10, 10, 3, 3,

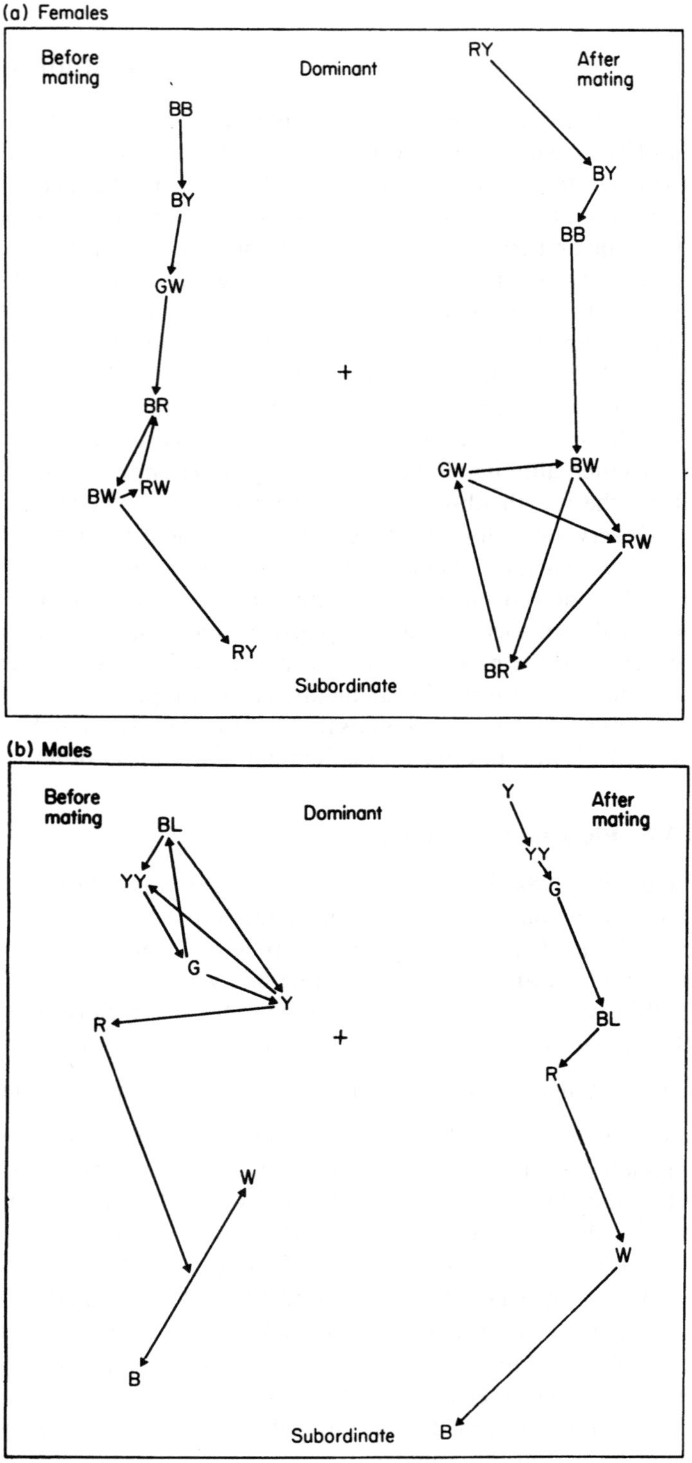

Fig. 6.9 Two-dimensional representation of pecking relationships for (a) female and (b) male pigeons (Table 6.6). For each sex the coordinates of the seven birds before and after mating were obtained from separate analyses of skew-symmetry, but they appear on the same graph for ease of comparison. Arrows indicate the direction of dominance. Axes have been rotated so that the maximum variation is in the vertical direction.

Table 6.7 Numbers of transitions between vegetation states in a heathland following regeneration after fire. (From Hobbs and Legg, 1984, with four minor states omitted.)

		Preceding state									
		B	L	G	E	EC	C	PC	P	CA	A
Succeeding state	B	15	0	0	0	0	4	0	0	0	0
	L	18	11	0	1	0	1	3	0	0	0
	G	47	17	5	0	0	0	1	0	0	0
	E	15	27	20	10	5	7	0	0	1	10
	EC	5	0	5	4	10	2	0	0	0	0
	C	1	8	8	21	5	18	0	3	1	21
	PC	1	1	1	3	4	11	101	7	0	0
	P	1	6	3	7	0	1	29	17	0	2
	CA	5	3	0	0	5	1	16	0	6	5
	A	3	14	8	0	0	3	3	5	9	7

Codes for vegetation states: B, bare ground; L, lichens (mostly *Cladonia* spp); G, grasses; E, *Erica cinerea*; C, *Calluna vulgaris*; P, pleurocarpous mosses (mostly *Hypnum jutlandicum*); A, *Arctostaphylos uva-ursi*. Two-letter codes indicate state with two species dominant.

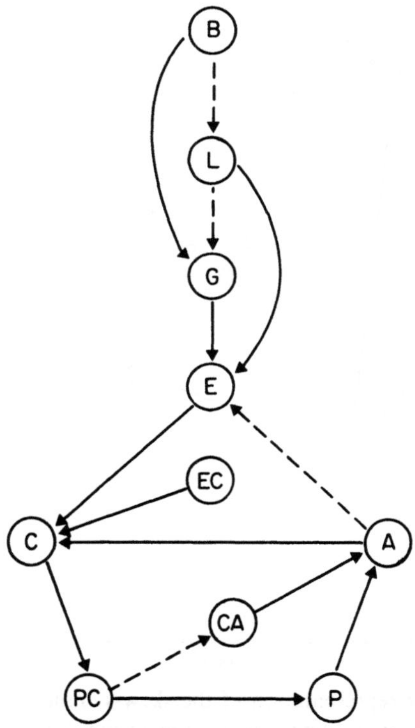

Fig. 6.10 Diagram indicating the main transitions between vegetation states (for codes see Table 6.7). The most likely transition from each state is indicated by a bold line, while other transitions with probability >0.15 are marked by dashed lines.

0.4, 0.4; therefore the first and second planes (pairs of axes), given in Figs 6.11 and 6.12 respectively, account for 77% and 16% of the variation in the skew-symmetric elements. The first plane (Fig. 6.11) shows the initial states of regeneration where positive transitions are indicated by moving anticlockwise about the origin and bold arrows indicate the major transitions from each state. The major transition from the state P_i is identified as being to the state P_j, where the triangle OP_iP_j is the largest that can be traversed in an anticlockwise direction. Thus for state B, OBG is the largest such triangle, while for state G it is OGE. Other transitions which also occur with high frequency are indicated by dashed lines. Transitions among states B, L, G and E are well represented by this figure. For example, states B, L and G lie almost on a straight line and $BL \simeq LG$. Hence the transition $B \rightarrow L$ is expected to be approximately equal to that from $L \rightarrow G$ and to be half that from $B \rightarrow G$. This is a good approximation to the observed transitions given in Table 6.7. Transitions among states close

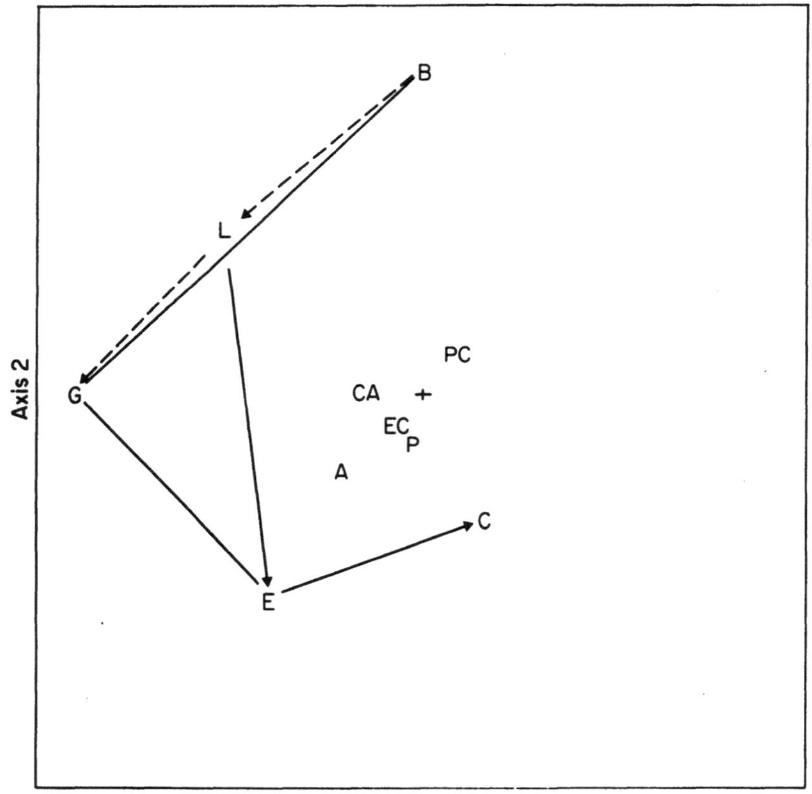

Fig. 6.11 First planar representation of the skew-symmetric part of the transition matrix for heathland regeneration (codes for vegetation states as in Table 6.7). Arrows indicate the major transitions shown in Fig. 6.10. + marks the origin.

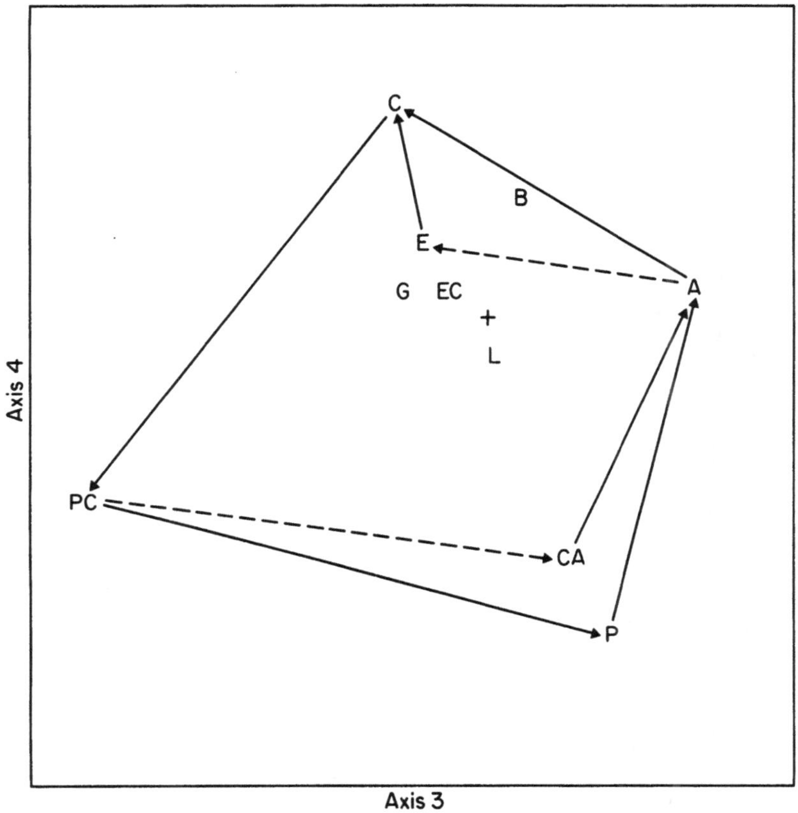

Fig. 6.12 **Second planar representation of the skew-symmetric part of the transition matrix for heathland regeneration (Table 6.7).**

to the origin are not well represented in this plane; for these one must look to dimensions three and four.

The second plane (Fig. 6.12) clearly shows the cyclical climax described earlier. The bold arrow from E to C is added from the first plane, while the other arrows come from the largest triangles, as above. Comparison with the skew-symmetric elements of Table 6.7 indicates that once again this triangular representation closely fits the observed transitions. This might be expected since the first two planes together account for 94% of the variation in skew symmetry. Moreover, the analysis has been exceptionally successful in separating the two stages of succession into orthogonal planes, the first representing the initial invasion of grasses, lichens followed by heather (*Erica*), the second a climax cycle of *Erica*, *Calluna*, *Hypnum* and *Arctostaphylos* species. This supplements the information from the classical Markovian analysis (Hobbs and Legg, 1984), which shows the changes in expected

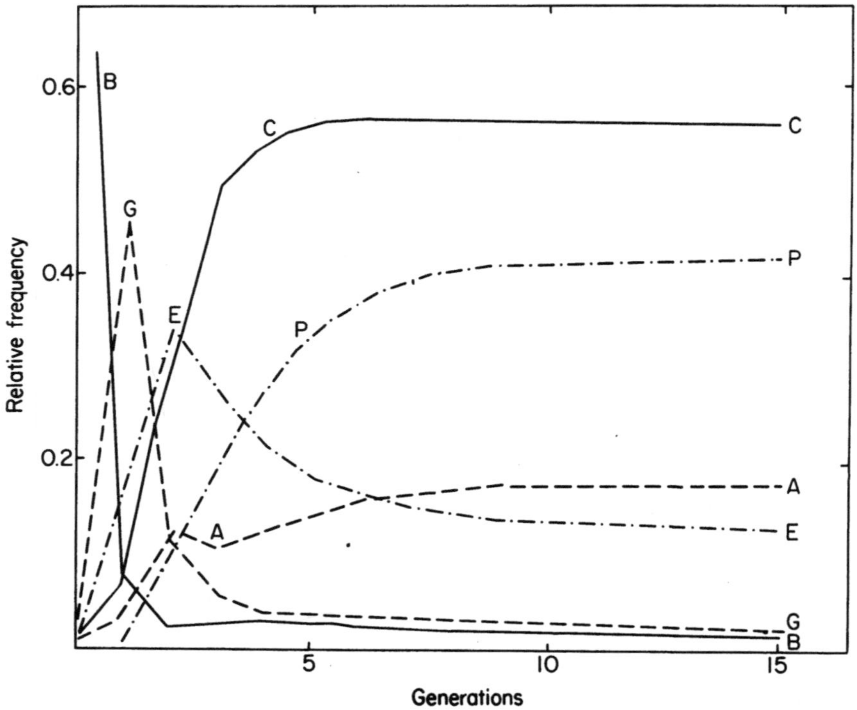

Fig. 6.13 Development of vegetation in a regenerating heathland using Markov model with transition probabilities between states derived from Table 6.7. (After Hobbs and Legg, 1984.)

frequency of species with time until an equilibrium state is reached (Fig. 6.13), but gives no indication of the dynamic nature of the composition of the vegetation at equilibrium.

6.3.3 Examining competitive interactions in plant mixtures

An experimental technique commonly used to study the competitive interactions between a number of plant species consists of growing each pair of species together in the same pot or plot and measuring the performance of each species separately. The results from such an experiment can be expressed as a square matrix with the (i, j)th element y_{ij} representing the mean performance (e.g. yield/half-pot) of the ith species when grown in association with the jth; the diagonal elements y_{ii} represent the comparable half-pot yields when each species is grown on its own.

Table 6.8 gives the individual species yields, expressed as $\log(\mathrm{g/m^2})$, in the first year of a mixture experiment involving nine grassland species (Jacquard

Table 6.8 Yields of nine grassland species when grown with different associate species. (From Jacquard and Caputa, 1970.)

	Species yield (log g/m^2)									
Associate species	*Li*	*Ae*	*Dg*	*Pp*	*Fp*	*Ms*	*Tp*	*Lc*	*Th*	*Mean yield of associate*
Lolium italicum	**9.36**	8.05	6.99	5.52	6.40	8.24	7.60	6.17	6.67	7.22
Arrhenatherum elatius	10.06	**9.04**	7.77	7.C6	6.97	9.20	8.63	7.18	7.65	8.17
Dactylis glomerata	10.10	9.39	**8.06**	6.66	8.29	9.28	8.59	7.47	7.90	8.41
Phleum pratense	9.95	9.24	8.29	**7.92**	8.24	9.25	8.83	8.22	7.98	8.66
Festuca pratensis	9.86	9.18	8.28	6.81	**7.85**	9.38	8.86	7.82	8.18	8.47
Medicago sativa	9.26	8.88	7.42	7.44	6.97	**9.15**	8.27	6.70	7.44	7.95
Trifolium pratense	9.82	9.09	7.42	7.39	7.56	8.97	**8.82**	6.80	7.48	8.15
Lotus corniculatus	10.02	9.29	8.33	8.44	8.10	9.15	8.96	**8.61**	8.22	8.79
Trifolium hybridum	9.77	9.35	8.45	8.30	8.12	9.19	8.97	8.00	**8.22**	8.70
Mean yield of species	9.80	9.06	7.89	7.28	7.61	9.09	8.61	7.44	7.75	

Bold figures indicate pure stand yields.

and Caputa, 1970). To help interpretation of the competitive interactions we first consider a biplot analysis of the log yields adjusted about their overall mean. The two-dimensional biplot shown in Fig. 6.14 describes 75% of the total variation in yield. The points for species and associates lie close to two intersecting straight lines and, recalling the diagnostic plots of Fig. 3.9, this suggests that the species log yields are well described by a model of the form

$$y_{ij}=\mu+g_i+h_j+\lambda g_i h_j, \qquad i=1, \ldots n; j=1, \ldots, n, \tag{6.4}$$

where g_i and h_j are respectively the direct and associate effects of the species, and λ represents the angular separation of the lines. This model has been discussed in the context of the competition diallel by Wright (1971).

It is of interest to compare equation (6.4) with a more general partitioning of yields for a competition diallel proposed by McGilchrist (1965). He expressed y_{ij} as the sum of three components

$$y_{ij}=y_{ii}+c_{ij}+z_{ij}, \tag{6.5}$$

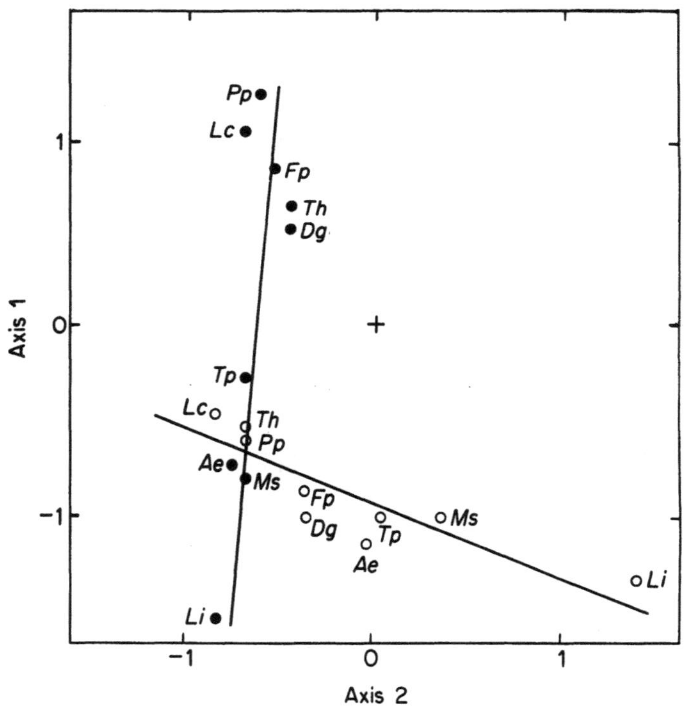

Fig. 6.14 Biplot analysis of centred species yields from a competition diallel analysis (Table 6.8). Each species is represented by a pair of points for its direct (solid circles) and its associate effect (open circles). The lines were obtained by fitting the model of equation (6.4) to the data.

where y_{ii} represents the pure stand yield;

$$c_{ij} = \tfrac{1}{2}(y_{ij} + y_{ji}) - \tfrac{1}{2}(y_{ii} + y_{jj}) \tag{6.6}$$

represents the combining ability of the mixture of species i and j, i.e. their excess yield as a mixture over the mean yield of their pure stands; and

$$z_{ij} = \tfrac{1}{2}(y_{ij} - y_{ii}) - \tfrac{1}{2}(y_{ji} - y_{jj}) \tag{6.7}$$

represents the competitive advantage of species i over species j. When y_{ij} represents the logarithm of yield, z_{ij} is identified with the logarithm of de Wit's relative crowding coefficient for species i with respect to species j (Harper, 1977). It will be noted that $c_{ij} = c_{ji}$ while $z_{ij} = -z_{ji}$, and hence the matrix $\mathbf{X}$ (with elements $x_{ij} = y_{ij} - y_{ii}$ representing the differential yields in mixtures compared with pure stands) can be partitioned into easily interpretable symmetric and skew-symmetric components which can be analysed separately. For the particular case where the yields are described by equation (6.4), the symmetric and skew-symmetric components are of the form

$$c_{ij} = -\tfrac{1}{2}\lambda(g_i - g_j)(h_i - h_j), \tag{6.8}$$

$$z_{ij} = (h_j - h_i)\{1 + \tfrac{1}{2}\lambda(g_i + g_j)\}. \tag{6.9}$$

In the general case, where the yields do not fit a simple model such as equation (6.4), the values c_{ij} may be identified with elements of an association matrix. A principal coordinates analysis then produces coordinates for the species in which species with relatively good combining ability appear close together. This is demonstrated for the nine grassland species in Fig. 6.15(a), where vectors are drawn from each species i to connect to that species j with which it has greatest combining ability (i.e. that $i \neq j$ for which c_{ij} is a maximum). Species occupying the centre of the plot, viz. *Dactylis*, *Festuca*, *Trifolium hybridum* and *Arrhenatherum*, can be identified as having good general combining ability.

An analysis of the skew-symmetric components of yield z_{ij} is shown in Fig. 6.15(b). Here the ordering of points is the same as that for the associate points on the biplot. Under an additive model of direct and associate effects, i.e. $\lambda = 0$ in equation (6.4), a linear relationship would be expected in Fig. 6.15(b) and distances between species points would then be proportional to the difference in their associate effects (from equation (6.9)).

The ordinations of species along the vertical axes in Figs 6.15(a) and (b) are very similar and suggest that species have greatest combining ability with species of similar competitive ability. The latter was in most cases closely related to pure stand yield (Wright, 1971).

6.4 A PROOF OF THE TRIANGLE-AREA THEOREM

We include here a proof of the result quoted earlier that $s_1(u_{i1}u_{j2} - u_{i2}u_{j1})$ is proportional to the area of the triangle with vertices at (u_{i1}, u_{i2}), (u_{j1}, u_{j2}) and

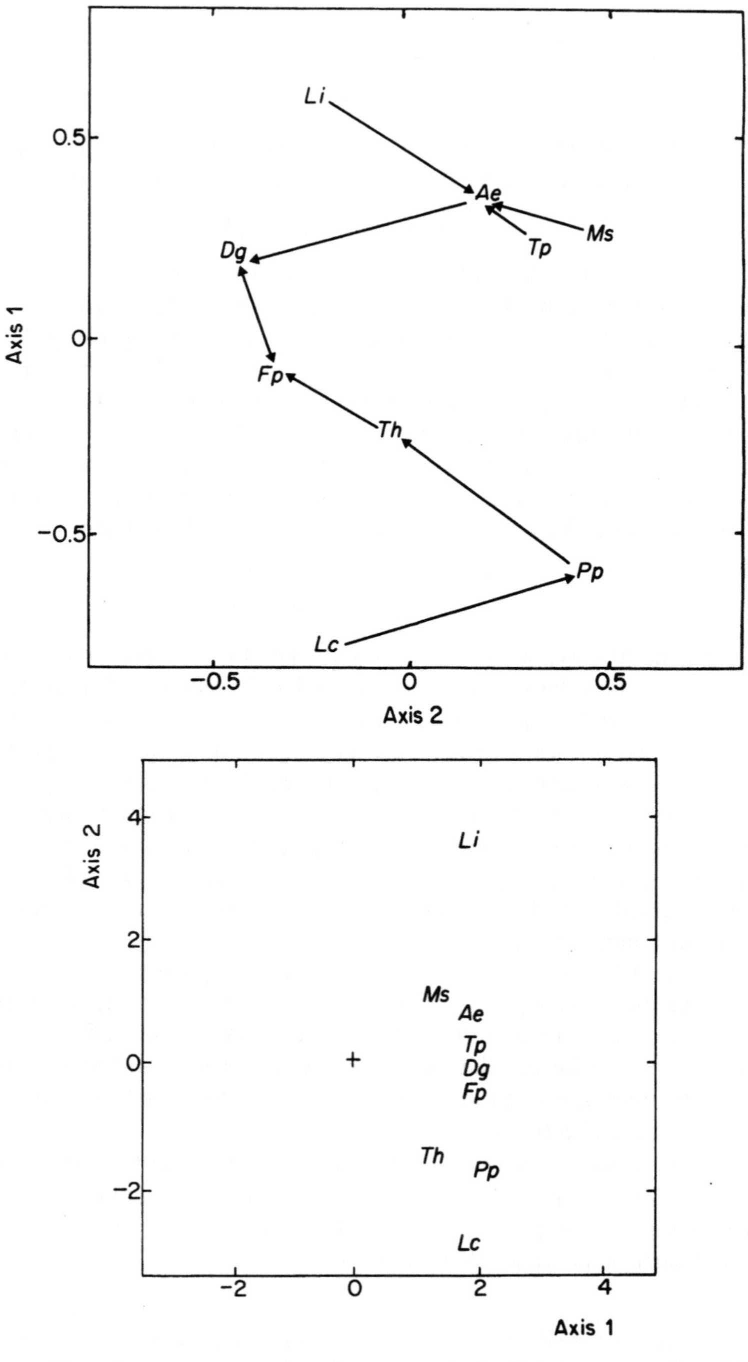

Fig. 6.15 First planar representations from an analysis of (a) the symmetric and (b) the skew-symmetric parts of a competition diallel matrix of yields (Table 6.8). In (a) each species is connected to the associate species with which it has greatest combining ability.

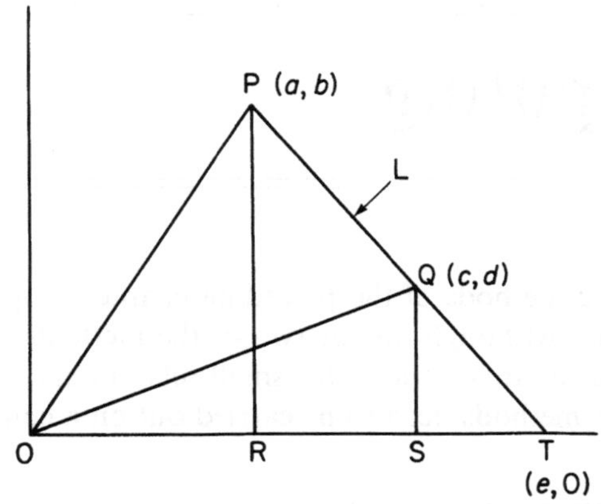

Fig. 6.16 Illustration of the triangle-area theorem.

the origin. Consider Fig. 6.16; we wish to show that the area of triangle OPQ is proportional to $bc-ad$.

Clearly the area of OPQ equals the area of OPT minus the area of OQT. Since the area of a triangle equals half its base times its height, the area of OPT is $\frac{1}{2}\times$(length of OT)$\times$(length of PR), i.e. $\frac{1}{2}eb$. Similarly, the area of OQT is $\frac{1}{2}ed$, thus the area of OPQ is $\frac{1}{2}e(b-d)$. To proceed we note that the equation of the line L is $(b-d)x+(c-a)y=bc-ad$; this can be verified by substituting both (a, b) and (c, d) for (x, y). Setting $y=0$ gives $(b-d)e=bc-ad$, from which the area of OPQ is simply $\frac{1}{2}(bc-ad)$, as required.

7 Computing

In practice, all the methods of the preceding chapters require some sort of computational aid. Although one or two of the methods can be used with programmable calculators, at least for small sets of data, the computations involved for most methods need to be carried out on a computer.

7.1 COMPUTING OPTIONS

There are four approaches that the reader might now adopt in respect of computer programming. Firstly, he may choose to write his own programs in a language such as Fortran, Pascal or Basic. Two other possibilities are to make use of existing programs written especially for ecologists, or those developed for specific multivariate methods. Lastly, the reader may choose to use one of the large general-purpose statistical packages. There are various advantages and disadvantages to each of these approaches which are discussed in the following sections. The various computer programs and packages that are mentioned in the text are referenced at the end of the chapter.

7.1.1 The 'do-it-yourself' approach

In some situations the approach of writing one's own programs is unavoidable; for example, the research worker in a developing country with no nearby computing centre and only a microcomputer has little choice but to write his own computer programs, although he may prefer to delay any formal data analysis until he has access to better computing facilities. Methods for preliminary inspection of the data, described in Chapter 2, may be useful in this case.

We hope that most, if not all, readers will choose not to write their own programs. The computational aspects may not be very difficult to encode in, say, Fortran, especially if subprograms from an existing library (e.g. the Numerical Algorithm Group subroutine library) are used for the matrix operations and decompositions. However, other parts of such programs are tedious to write and are often extremely lengthy; for example, producing graphical output, either on a line printer or on a plotter. There are the additional problems that mistakes can be made and, less obviously, numerical instability can occur, both of which will result in incorrect analyses.

Numerical instability occurs because computers are unable to store numbers with complete accuracy. Typically, individual values may be stored with their first seven, or sometimes fifteen, digits correct; however, there can be a large loss of accuracy when numbers are combined arithmetically. A well-known example concerns the calculation of the variance of a set of values: if $\bar{x}$ is the mean of n values x_i ($i=1, 2, \ldots, n$), the two formulae for the variance

$$\frac{1}{n}\sum_{i=1}^{n}(x_i-\bar{x})^2 \quad \text{and} \quad \frac{1}{n}\sum_{i=1}^{n}x_i^2-\bar{x}^2$$

are algebraically equivalent: however, if these two forms are used on a computer they can give wildly differing results, especially when the individual values of x_i are large; indeed, the second formula can then easily give a negative 'variance'.

If computer programs are to be written from scratch, the choice of computer language is important. Basic is a popular, relatively simple language, and many computers can run Basic programs. However, there are many versions of Basic, with major differences between them, which can lead to problems when a program is moved from one computer to another. For the sort of extensive programming needed for multivariate analysis we cannot recommend Basic. Pascal is an excellent language which is widely available; also, it suffers less from having a variety of versions. Unfortunately, the 'standard' for the language lacks double-precision arithmetic, which is necessary for some calculations (e.g. forming sums of squares). We recommend that, when possible, Fortran 77 be used, or a version of Pascal with high-precision arithmetic.

7.1.2 Programs for ecologists

A number of computer programs have been designed specifically for use by ecologists. Among these the Cornell Ecology Programs (Gauch, 1982) seem to be established as the most widely used package. They consist of a suite of independent programs for the most common ordination and classification methods, and also provide facilities for general data manipulation before and after more formal analysis. Two of the programs are for methods whose initial development was in the form of these computer algorithms, viz. DECORANA, for detrended correspondence analysis (Section 3.6.2), and TWINSPAN for two-way indicator species analysis (Section 5.2). One particular advantage of such programs is that they are usually designed to handle large data sets, which can be a stumbling block for some of the general-purpose packages described later. The fact that such programs are written for, and largely by, ecologists generally makes them easier to use by the novice. In particular, the options should be restricted to those of ecological relevance and the output should be in a form most likely to be useful to the ecologist, and appropriately annotated. However, such tailor-made programs necessarily suffer from inflexibility and the options available to the user are often governed by the current fashion or fads of the programmer.

Kershaw and Looney (1985) argue that the Cornell Ecology Programs provide all the necessary methods for multivariate analysis in ecology. While we would agree that it is unnecessary, and indeed only confusing, to provide absolutely all possible variants of ordination and classification methods, the Cornell Ecology Programs do not currently include some important techniques described in this book, such as those of Chapters 4 and 6.

7.1.3 Programs for multivariate methods

Some multivariate methods are only available in specialized computer programs. A good example until recently is non-metric ordination, for which many such programs exist. Schiffman, Reynolds and Young (1981) discuss a number of these, including KYST which we used in Chapter 3. Everitt (1980) lists a number of classification programs. In recent years, various programs have been written in France that handle large data sets; although intended mainly for sociological studies, they could be used for large sets of ecological data: Jambu and Lebeaux (1983) describe some of these programs as well as giving listings of Fortran code.

In general, such special-purpose programs can prove useful: the popular ones have been used often enough for mistakes to have been removed; the generality of their application means that most of the useful output will be given, although this may need to be edited or further processed (see Section 7.2).

7.1.4 General-purpose packages

Finally, there are a number of large, general-purpose, statistical packages, e.g. BMDP (which is a unified collection of programs), Genstat, SAS and SPSS. All of these provide some multivariate methods, though none has all the methods described in this book directly available.

Genstat is the most powerful of these packages because it allows the user to prepare a computer program in a high-level statistical language. This provides a flexibility of use that is absent from the other three, with the possible exception of SAS. A completely revised version of Genstat (Genstat 5) is now available: this is considerably easier to use than earlier versions. Most of the methods of analysis that we have described are directly available in Genstat (Table 7.1) and others are available by using pre-written sets of instructions (procedures) in the high-level language; a library of such procedures is available at every site that has Genstat. One particular advantage of Genstat is that it provides all the usual operations of matrix algebra, so that any of the methods described in terms of matrices can be programmed very easily. We

Table 7.1 Some useful Genstat directives for multivariate analysis.

Directive	*Purpose*
FLRV	Computes the latent roots, latent vectors and trace of a symmetric matrix.
SVD	Computes the singular value decomposition of a rectangular matrix.
PCP	Computes the principal components analysis of a set of variates giving the principal components loadings, associated roots, principal components scores, and the residuals (if fitting less than all the components).
CVA	Computes the canonical variate analysis of a set of variates for a specified grouping of the units, giving the canonical variate loadings, associated roots, canonical variate means, and the residuals (if fitting less than all the canonical variates).
PCO	Computes the principal coordinates analysis of a given similarity matrix. Principal coordinates, associated roots, distances from the centroid and residuals are produced. Principal components and canonical variate analysis can also be computed using this directive.
ROTATE	Rotates a configuration of points to fit best another configuration, so as to minimize the sum of squared distances between corresponding points of the two configurations.
CLUSTER	Obtains the non-hierarchical classification of a number of units so as to optimize one of several possible criteria calculated from variate values. The criteria include sums of squares between classes, maximal prediction, and Mahalanobis-D^2 between classes.
FSIMILARITY	Provides several facilities for forming, summarizing and printing similarity matrices.
HCLUSTER	Computes a hierarchical cluster analysis according to one of five possible clustering methods. Also produces dendrograms.
HDISPLAY	Displays minimum spanning tree and nearest neighbours of each unit.

have used Genstat to prepare nearly all the analyses and examples presented in this book.

A disadvantage of some of these packages, including Genstat, is that they cannot handle very large data sets, although this problem can sometimes be avoided by handling the data in manageable blocks. This is what we had to do in constructing the similarity matrices for the Rothamsted Insect Survey sites based on the abundances of 510 species (Section 4.3). An alternative approach when analysing large data sets is to take advantage of the properties of matrix algebra to reduce a large problem to a smaller one, which is shown in the second example in the following section.

7.2 EXAMPLES OF GENSTAT PROGRAMS

In this section we describe two input programs, and one procedure, to give potential users some flavour of Genstat programming in the new version, Genstat 5. These programs, together with the data in some other file, are what the user would provide as input to Genstat.

The first example (Table 7.2) shows how principal components analysis and correspondence analysis can be used in Genstat. The first statement serves to identify the job; the text within apostrophes on line 2 is simply a comment, and Genstat will ignore it. On line 3 we declare two structures to contain codes for the rows and columns of the data matrix. These codes will be in the form of text, and we can use the identifiers, labels and columns, to refer to them later in the program. The statement on line 4 will read the column labels; the options, within square brackets, specify that these codes are in a secondary input file (channel), and that Genstat is to continue reading codes until it finds an end-of-data marker, i.e. the number of values of columns is to be set by the read statement. Line 5 declares two scalars which will be set to the numbers of rows and columns; the latter is obtained on line 7 (calc is short for calculate; however Genstat only requires the first four letters of statement names). Line 8 specifies a single identifier v which will be used to represent the data matrix as a whole; it does this by pointing to a set of structures which will hold each column of the data matrix. Together with the row labels, this set of structures is input on line 9, and printed on line 11. Lines 13 and 14 declare a matrix to hold the principal components scores, and a single structure to hold the latent roots (eigenvalues) and vectors from the analysis. Line 16 specifies the principal components analysis: the print option tells Genstat what output is required; the input is taken from the structure v, and the latent roots and vectors, and the scores, are stored for subsequent plotting. A biplot of the scores and latent vectors is requested by the statement on line 17; in fact there is no direct statement in Genstat for this, and line 17 is a procedure call – the procedure is given in Table 7.3 and described later in this section.

The first part of this program is relatively straightforward, since principal components analysis is directly available in Genstat. Correspondence analysis is not directly available, i.e. there is no single statement within the Genstat language to do it, although it is available as a standard Genstat procedure. We could have called that procedure within this program; however, it is more instructive to show how the analysis is specified using matrix operations (which is what the procedure does, anyway). It is best to calculate the correspondence analysis results for min(n, p) dimensions: this quantity is obtained on lines 19 and 20, and then used in the declarations (lines 21–28) of

Table 7.2 Genstat 5 input program for two multivariate analyses. The line numbers are included here for ease of reference, but should be omitted when used as input to Genstat.

```
Job        'PCP/CA'
      " labels for rows and columns "
text       labels,cols
read       [channel=2; setnval=yes] cols
scalar     n,p
      " define data, and read it "
calc       p = nval(cols)
pointer    [nval=cols] v
read       [channel=2; setnval=yes] labels,v[1...p]
calc       n = nval(labels)
print      labels,v[1...p]
      " structures for keeping results "
matrix     [labels; p] score
lrv        [cols; p]   lrv
      " principal components analysis "
pcp        [print=loadings,roots,scores] v; lrv=lrv; scores=score
bigraph    [ndim=3] score; lrv['Vectors']; labels; cols
      " define data structures for correspondence analysis "
scalar     minnp
calc       minnp = vmin(n,p)
matrix     [labels; cols]  x
  &        [labels; minnp] a
  &        [cols;   minnp] b
  &        [labels; 1]     onen,rv
  &        [cols;   1]     onep,cv
diagonal   [labels]        r
  &        [cols]          c
  &        [minnp]         s,%s
      " correspondence analysis via matrix operations "
calc       x $ [!(1...n); 1...p] = v[1...p]
  &        onen,onep = 1
  &        rv = product(x; onep)
  &        cv = ltproduct(x; onen)
equate     rv,cv; r,c
calc       [zdz=zero] r = sqrt(r / r / r)
  &                   c = sqrt(c / c / c)
calc       x = r *+ x *+ c
svd        x; a; s; b
calc       a = r *+ a
  &        b = c *+ b
  &        s = s * s
  &        %s = 100 * s / (sum(s) - 1)
      " display results "
print      s,%s,a,b; fieldwidth=10; decimal=5,2,5,5
bigraph    [col1=2] a; b; labels; cols
endjob
stop
```

the various matrices needed for the analysis. We start by first copying the data matrix into the matrix x on line 30: this is done by copying each complete column, one at a time. On line 31 the values of the column vectors onen and onep are set to one. These are then used on lines 32 and 33 to calculate the row and column sums of x; the function product gives the matrix product, and ltproduct also gives a matrix product, but with the left matrix (here x) transposed. Line 34 simply copies the values of rv and cv into the diagonal matrices r and c. The calculations on lines 35 and 36 replace r and c by their inverse square-root matrices; the convolution of these statements, together with the option specifying that zero-divided-by-zero is to be taken as zero, copes with zero row or column sums. Line 37 scales the data matrix by the inverse square-root matrices: the *+ symbols provide a convenient, alternative, shorthand for matrix products. Line 38 requests the singular value decomposition of x into the matrices a, s, and b. The matrices of correspondence analysis scores are obtained on lines 39 and 40. Lines 41 and 42 square the singular values, and then express them as percentages of their total, subtracting 1 which comes from the trivial solution. Lastly the results are printed and plotted.

In the program of Table 7.2 we used the procedure bigraph to produce plots of points for rows and columns, i.e. biplots and correspondence analysis plots. This procedure is listed in Table 7.3. Line 1 introduces the procedure and gives its name. After the comment on line 2, lines 3 and 4 specify the options and parameters used when calling the procedure. There are two options: the first,

Table 7.3 Genstat 5 procedure bigraph.

```
1    procedure 'bigraph'
2           " define options and parameters "
3    option     'NDIM','COL1'; mode=p; default=2,1
4    parameter 'A','B','ALAB','BLAB'; mode=p
5           " get sizes "
6      scalar     n,p,r
7      calc       n,p = nrow(A,B)
8        &        r = NDIM + COL1 - 1
9           " get variates to plot "
10     variate    [n] av[1...NDIM]
11       &        [p] bv[1...NDIM]
12     calc       av[1...NDIM] = A $ [!(1...n); COL1...r]
13       &        bv[1...NDIM] = B $ [!(1...p); COL1...r]
14          " loops over dimensions "
15     for        ya = av[2...NDIM]; yb = bv[2...NDIM]
16       for      xa = av[1...NDIM]; xb = bv[1...NDIM]
17          " exit for irrelevant graphs "
18         exit  xa .is. ya
19          " draw graph "
20         graph [equal=scale; 61; 101] ya,yb; xa,xb; symbols=ALAB,BLAB
21       endfor
22     endfor
23   endprocedure
```

called NDIM, indicates how many dimensions are to be plotted; the second, called COL1, indicates which column of the matrices of scores is to be the first dimension. If the user does not specify values for the options, they will take the default values 2 and 1, respectively. Of the four parameters, A and B are for the matrices of row and column scores, and ALAB and BLAB are for the row and column labels. The mode=p part of each specification is required for a technical reason which need not concern us here. The procedure is designed to plot graphs of the scores for all pairwise combinations of dimensions within the requested range of columns. For example, in the first call of the procedure on line 17 of Table 7.2, the first option (NDIM) is set to 3 and COL1 takes its default value of 1: this produces graphs of the scores from columns 2 v. 1, 3 v. 1, and 3 v. 2. The second option has been introduced to cope with matrices of scores from correspondence analysis, so that in the second call (line 45) COL1 has been set to 2 and a single plot is produced of the scores from columns 3 v. 2.

Lines 6–8 of Table 7.3 calculate three useful size quantities, which are used in lines 12–13 in extracting the relevant columns from the matrices of scores into sets of variates declared on lines 10–11. Line 15 controls the looping over the y variates for each graph, so that on the first pass ya becomes av[2] and yb becomes bv[2], and so on. Likewise, line 16 controls the looping over the x variates. Without any extra control on the looping, this would produce many more graphs than are actually necessary, e.g. av[2] will be plotted against av[2], and a plot of av[2] against av[3] will be produced as well as the plot of av[3] against av[2]. However, if we can arrange that execution of the inner loop is finished as soon as xa becomes the same as ya, these redundant graphs will not be plotted: this is exactly the effect of line 18. The graphs are drawn by line 20. The early part of the statement specifies: (i) that the x- and y-axes are to be on equal scales; (ii) that 31 line printer rows are to make up the graph frame; (iii) that 51 line printer columns are to be used for the frame. Thus the graph will be mathematically square and will be drawn within a physical square of side 5 inches: this is important, and should be borne in mind when graphs are produced. (The numbers of rows and columns for the graph frame that are mentioned here will be correct for printers set up to print the standard 10 characters per inch and 6 rows per inch.) The final three lines of the procedure are the ends of the two nested loops, and the procedure itself.

The second example relates directly to an example discussed in Section 4.2. Here we have fourteen sites, coded A to N, and 510 species; we are interested in an ordination of the sites using principal components analysis. This data set is of moderate size and could prove troublesome for some general-purpose packages: for example a conventional approach would use 510 variates of length 14 and attempt to find the latent roots and vectors of the sum of squares and products (SSP) matrix of order 510. However, knowing certain properties of matrix algebra (Section 3.2.1) we can proceed as below.

If the data are represented by the matrix $\mathbf{X}$ (14×510) then we know that the matrix of principal components scores $\mathbf{A}$ can be obtained by first column-centring $\mathbf{X}$ and then finding its singular value decomposition (SVD). If $\mathbf{H}$ is the centring matrix of order 14, we have the centred form of $\mathbf{X}$ as $\mathbf{Y}=\mathbf{HX}$, and then from the SVD, $\mathbf{Y}=\mathbf{USV}'$, we get $\mathbf{A}=\mathbf{US}$. The usual computational approach would form $\mathbf{Y}'\mathbf{Y}$, the SSP matrix of order 510, and then obtain the spectral decomposition $\mathbf{Y}'\mathbf{Y}=\mathbf{VS}^2\mathbf{V}'$ followed by the scores $\mathbf{A}=\mathbf{YV}(=\mathbf{US})$ (see Appendix for the relationship between singular value and spectral decompositions). However, it is far easier here to operate with $\mathbf{YY}'$ which is of order 14. We know that $\mathbf{YY}'=\mathbf{US}^2\mathbf{U}'$, so we can easily obtain the scores $\mathbf{US}$ from the spectral decomposition of $\mathbf{YY}'$. (In fact all that this amounts to is the spectral decomposition of $\mathbf{HXX}'\mathbf{H}'$, which corresponds to principal coordinates analysis, Section 3.5.)

The solution to this can be programmed quite easily, and is given in Table 7.4. Most of the statements have already been used in Tables 7.2 and 7.3; however, two points are worth mentioning. Firstly, in centring x on line 12, the matrix products have been arranged so that the square matrix $11'$, of order n, is not explicitly formed, which it would have been with the statement calc x = x − product (rtproduct (one; one)/14; x). Secondly, in the declaration of the structure to hold the latent roots and vectors on line 16, the matrix for the vectors and the diagonal matrix for the roots have been named explicitly; the latent roots and vectors are actually formed on line 17.

Table 7.4 Genstat 5 input program for analysis of a large data matrix.

```
 1   Job         'Principal components analysis via
 2                transposed spectral decomposition'
 3            " labels for the 14 sites "
 4   text        [14] sites
 5   read        [channel=2] sites
 6            " data matrix "
 7   matrix      [sites; 510] x
 8   read        [channel=2] x
 9            " centre x "
10   matrix      [14; 1] one
11   calc        one = 1
12     &         x = x - product( one; ltproduct( one; x) / 14 )
13            " transposed spectral decomposition "
14   symmetric [sites] xxt
15   calc        xxt = rtproduct( x; x)
16   lrv         [sites; 14] l; vectors=a; roots=r
17   flrv        xxt; l
18            " calculate scores and print results "
19   calc        r = r * (r > 0)
20     &         a = a *+ sqrt(r)
21   print       r,a
22   endjob
23   stop
```

7.3 HANDLING MISSING VALUES

In the statisticians' ideal world there would be no missing values; unfortunately this is not matched in ecological reality. Standard software sometimes deals with missing values, but not always in the optimum way. For example, in Genstat, the statement used for principal components analysis will omit all units for which any values are missing; however, a Euclidean similarity matrix can be calculated (see Section 1.4.3) and used as input to principal coordinates analysis (see Section 3.5.3) – here full use is made of the data that are present. Some of the specialized programs for multivariate analysis cope with missing values, e.g. KYST does so, and this is a good case for its use.

For specific types of analysis it is sometimes possible to construct an algorithm that will take into account the fact that some data are missing. Everitt and Gower (1981) describe an example where a weighted generalized Procrustes method was required to handle a data set much of which was absent. Another example is that of Digby (1979) where a particular multiplicative model was fitted iteratively, using a Genstat program. This was a special case of a more general multiplicative model fitting procedure for use with missing data, described by Gabriel and Zamir (1978).

7.4 CONCLUSION

We strongly recommend that readers do not start writing their own programs in a computer language such as Fortran or Basic. It is far easier, and much quicker, to write programs using a standard computer package.

Rather than using specialized programs, we recommend Genstat: we use it on a day-to-day basis and find very few occasions when another program is needed. However, the use of specialized programs is not totally avoidable; for example, we have used KYST for some of the examples in this book, and in other cases the analysis may have been more straightforward if we had used a multivariate program specifically written to process large data sets. However, even then we have found Genstat to be useful both in preparing the input data for the other program, and in displaying the results of the analyses.

7.5 LIST OF SOFTWARE

Further details of the software mentioned in this chapter may be obtained from the addresses listed below.

BMDP	BMDP Statistical Software 1964 Westwood Boulevard Suit 202 Los Angeles California 90025 USA

Cornell Ecology Programs	H. G. Gauch The Editor, Cornell Ecology Programs Cornell University Ithaca New York USA
Genstat	The Statistical Package Coordinator Numerical Algorithms Group 256 Banbury Road Oxford OX2 7DE UK
KYST	The Supervisor Bell Labs Computing Information Library 600 Mountain Avenue Murray Hill New Jersey 07974 USA
NAG	Numerical Algorithms Group 256 Banbury Road Oxford OX2 7DE UK
SAS	SAS Institute Inc. Cary North Carolina USA
SPSS	SPSS Inc. 444 N. Michigan Avenue Chicago IL 60611 USA

References

Banfield, C. F. (1978). Singular value decomposition in multivariate analysis. In *Proceedings of the Institute of Mathematics and its Applications on Numerical Software – Needs and Availability* (ed. D. Jacobs), pp. 137–49. Academic Press, London.

Benzecri, P. J. (1973). L'analyse des correspondances. *L'analyse des Données*, Vol. 2. Dunod, Paris.

Blackith, R. E. and Reyment, R A. (1971). *Multivariate Morphometrics*. Academic Press, London.

Bradu, D. and Gabriel, K. R. (1978). The biplot as a diagnostic tool for models of two-way tables. *Technometrics* **20**: 47–68.

Bray, J. R. and Curtis, J. T. (1957). An ordination of the upland forest communities of southern Wisconsin. *Ecological Monographs* **27**: 325–49.

Brown, J. L. (1975). *The Evolution of Behaviour*. Norton, New York.

Brown, R. T. and Curtis, T. T. (1952). The upland conifer-hardwood forests of northern Wisconsin. *Ecological Monographs* **22**: 217–34.

Cain, A. J. and Harrison, G. A. (1958). An analysis of the taxonomists' judgement of affinity. *Proceedings of the Zoological Society of London* **131**: 85–98.

Clifford, H. T. and Stephenson, W. (1975). *An Introduction to Numerical Classification*. Academic Press, New York.

Constantine, A. G. and Gower, J. C. (1982). Models for the analysis of interregional migration. *Environment and Planning A* **14**: 477–97.

Coppock, J. T. (1976). *Agricultural Atlas of England and Wales*. Faber and Faber, London.

Cormack, R. M. (1971). A review of classification. *Journal of the Royal Statistical Society*, Series A, **134**: 321–67.

Czekanowski, J. (1909), Zur differential diagnose der Neandertalgruppe. *Korrespbl. dt. Ges. Antwop.* **40**: 44–7.

Davies, R. G. (1984). Some problems of numerical taxonomy. *Science Progress* **69**: 315–39.

Delany, M. J. and Healy, M. J. R. (1964). Variations in the long-tailed field mouse (*Apodemus sylvaticus L.*) in north-west Scotland. II: Simultaneous examination of all characters. *Proceedings of the Royal Society of London, B* **161**: 200–7.

Digby, P. G. N. (1979). Modified joint regression analysis for incomplete variety × environment data. *Journal of Agricultural Science* **93**: 81–6.

Diggle, P. J. (1983). *Statistical Analysis of Spatial Point Patterns*. Academic Press, London.

Eckart, C. and Young, G. (1936). The approximation of one matrix by another of lower rank. *Psychometrika* **1**: 211–318.

Edwards, A. W. F. (1972). *Foundations of Mathematical Genetics*. Cambridge University Press.

Ehrenburg, A. S. C. (1982). *A Primer in Data Reduction*. Wiley, New York.

Ellenberg, H. (1979). Zeigerwerte der Gëfasspflanzen Mitteleuropas. *Scripta Geobotanica*, Vol. 9 (2nd edn). Goltze, Gottingen.

Everitt, B. S. (1980). *Cluster Analysis* (2nd edn). Heinemann Educational Books, London.

Everitt, B. S. and Gower, J. C. (1981). Plotting the Optimum Positions of an Array of Cortical Electrical Phosphenes. In *Interpreting Multivariate Data* (ed. V. Barnett), pp. 279–87. Wiley, Chichester.

Fisher, R. A. and Yates, F. (1963). *Statistical Tables for Biological, Agricultural and Medical Research*. Oliver and Boyd, Edinburgh.

Gabriel, K. R. (1971). The biplot graphic display of matrices with application to principal component analysis. *Biometrika* **58**: 453–67.

Gabriel, K. R. (1981). Biplot display of multivariate matrices for inspection of data and diagnosis. In *Interpreting Multivariate Data* (ed. V. Barnett), pp. 147–73. Wiley, Chichester.

Gabriel, K. R. and Zamir, S. (1978). Lower rank approximation of matrices by least squares with any choice of weights. *Technometrics* **21**: 489–98.

Gauch, H. G. (1982). *Multivariate Analysis in Community Ecology*. Cambridge University Press.

Gauch, H. G., Whittaker, R. H. and Singer, S. B. (1981). A comparative study of non-metric ordinations. *Journal of Ecology* **69**: 135–52.

Gauch, H. G., Whittaker, R. H. and Wentworth, T. R. (1977). A comparative study of reciprocal averaging and other ordination techniques. *Journal of Ecology* **65**: 157–74.

Gittins, R. (1979). Ecological application of canonical analysis. In *Multivariate Methods in Ecological Work* (eds L. Orlóci, C. R. Rao and W. M. Stiteler), pp. 309–535. International Cooperative, Burtonsville, Md.

Goodall, D. W. (1973). Sample similarity and species correlation. In *Ordination and Classification of Communities*, Part V: *Handbook of Vegetation Science* (ed. R. H. Whittaker), pp. 105–56. W. Junk, New York.

Gordon, A. D. (1981). *Classification*. Chapman and Hall, London.

Gower, J. C. (1966a). Some distance properties of latent root and vector methods used in multivariate analysis. *Biometrika* **53**: 325–38.

Gower, J. C. (1966b). A Q-technique for the calculation of canonical variates. *Biometrika* **53**: 588–9.

Gower, J. C. (1971a). A general coefficient of similarity and some of its properties. *Biometrics* **27**: 857–72.

Gower, J. C. (1971b). Statistical methods of comparing different multivariate analyses of the same data. In *Mathematics in the Archaeological and Historical Sciences* (eds F. R. Hodson, D. G. Kendall and P. Tautu), pp. 138–49. Edinburgh University Press, Edinburgh.

Gower, J. C. (1974). Maximal predictive classification. *Biometrics* **30**: 643–54.

Gower, J. C. (1975). Generalised Procrustes analysis. *Psychometrika* **40**: 33–51.

Gower, J. C. (1977). The analysis of asymmetry and orthogonality. In *Recent Developments in Statistics* (eds J. Barra, F. Brodeau, G. Romier and B. van Cutsen), pp. 109–23. North Holland, Amsterdam.

Gower, J. C. (1985). Measures of similarity, dissimilarity and distance. In *Encyclopaedia of Statistics*, Vol. 5 (eds N. L. Johnson, S. Kotz and C. B. Read), pp. 397–405. Wiley, New York.

Gower, J. C. and Digby, P. G. N. (1981). Expressing complex relationships in two dimensions. In *Interpreting Multivariate Data* (ed. V. Barnett), pp. 83–118. Wiley, Chichester.

Gower, J. C. and Legendre, P. (1986). Metric and Euclidean properties of dissimilarity coefficients. *Journal of Classification* **3**: 5–48.

Gower, J. C. and Ross, G. J. S. (1969). Minimum spanning trees and single linkage cluster analysis. *Applied Statistics* **18**: 54–64.

Graybill, F. A. (1969). *Introduction to Matrices with Applications in Statistics.* Wadsworth, Belmont, CA.

Green, P. E. and Carroll, J. D. (1976). *Mathematical Tools for Applied Multivariate Analysis.* Academic Press, New York.

Green, R. H. (1979). *Sampling Design and Statistical Methods for Environmental Biologists.* Wiley, New York.

Greenacre, M. J. (1984). *Theory and Applications of Correspondence Analysis.* Academic Press, London.

Greig-Smith, P. (1983). *Quantitative Plant Ecology* (3rd edn). Blackwell Scientific Publications, Oxford.

Guhl, A. M. (1953). Social behaviour of the domestic fowl. *Kansas Agricultural Experimental Station, Technical Bulletin,* Vol. 73.

Guhl, A. M. (1956). The social order of chickens. *Scientific American* **194**: 42–6.

Harper, J. L. (1977). *Population Biology of Plants.* Academic Press, London.

Healy, M. J. R. and Taylor, L. R. (1962). Tables for power-law transformations. *Biometrika* **49**: 557–9.

Hill, M. O. (1973). Reciprocal averaging: an eigenvector method of ordination. *Journal of Ecology* **61**: 237–49.

Hill, M. O. (1979a). *DECORANA – A Fortran program for detrended correspondence analysis and reciprocal averaging.* Cornell University, Ithaca, New York.

Hill, M. O. (1979b). *TWINSPAN – A Fortran program for arranging multivariate data in an ordered two-way table by classification of the individuals and attributes.* Cornell University, Ithaca, New York.

Hill, M. O., Bunce, R. G. H. and Shaw, M. W. (1975). Indicator species analysis, a divisive polythetic method of classification, and its application to a survey of native pinewoods in Scotland. *Journal of Ecology* **63**: 597–613.

Hill, M. O. and Gauch, H. G. (1980). Detrended correspondence analysis, an improved ordination technique. *Vegetatio* **42**: 47–58.

Hobbs, R. J. and Legg, C. J. (1984). Markov models and initial floristic composition in heathland vegetation dynamics. *Vegetatio* **56**: 31–43.

Ivimey-Cook, R. B. and Proctor, M. C. F. (1966). The application of association analysis to phytosociology. *Journal of Ecology* **54**: 179–92.

Jaccard, P. (1901). Distribution de la flore alpine dans le Bassin des Dranses et dans quelques régions voisines. *Bull. Soc. vand. Sci. Nat.* **37**: 241–72.

Jacquard, P. and Caputa, J. (1970). Comparaison de trois modèles d'analyse des relations sociales entre espèces végétales. *Annales de l'Amelioration des Plantes* **20**: 115–58.

Jambu, M. and Lebeaux, M. O. (1983). *Cluster Analysis and Data Analysis.* North Holland, Amsterdam.

Kempton, R. A. (1979). The structure of species abundance and measurements of diversity. *Biometrics* **35**: 307–21.

Kempton, R. A. (1981). The stability of site ordinations in ecological surveys. In *The Mathematical Theory of the Dynamics of Biological Populations* (eds R. W. Hiorns and D. Cooke), pp. 217–30. Academic Press, London.

Kempton, R. A. (1984). The use of biplots in interpreting genotype by environment interactions. *Journal of Agricultural Science* **103**: 123–35.

Kershaw, K. A. and Looney, J. H. H. (1985). *Quantitative and Dynamic Plant Ecology*. Edward Arnold, London.

Lance, G. N. and Williams, W. T. (1966). Computer programs for hierarchical polythetic classification. *Computer Journal* **9**: 60–4.

Lance, G. N. and Williams, W. T. (1968). Note on a new information-statistic classificatory program. *Computer Journal* **11**: 195.

Legendre, L. and Legendre, P. (1983). *Numerical Ecology*. Elsevier, Amsterdam.

Lowe, H. J. B. (1984). The assessment of populations of the aphid *Sitobion avenae* in field trials. *Journal of Agricultural Science* **102**: 487–97.

McGilchrist, C. A. (1965). Analysis of competition experiments. *Biometrics* **21**: 975–86.

Mardia, K. V., Kent, J. T. and Bibby, J. M. (1979). *Multivariate Analysis*. Academic Press, London.

Margalef, T. (1958). Distribución de los crustaceos en las aguas continentales españolas. Grado de asociación entre las especies en relación con factores ecológicos e históricos. *Publnes Inst. Biol. apl., Barcelona* **27**: 17–31.

Masure, R. H. and Allee, W. C. (1934). The social order in flocks of the common chicken and pigeon. *Auk* **51**: 306–25.

Mitchley, J. and Guarino, L. (1984). Canonical analysis of asymmetric matrices: two applications from a study of chalk grassland in the south of England. *Vegetatio* **57**: 53–60.

Mountford, M. D. (1962). An index of similarity and its application to classificatory problems. In *Progress in Soil Zoology* (ed. P. W. Murphy), pp. 43–50. Butterworths, London.

Mourant, A. E., Kopéc, A. C. and Domariewska-Sobezak, K. (1976). *The Distribution of the Human Blood Groups and other Polymorphisms* (2nd edn). Oxford University Press, London.

Mozley, A. (1936). The statistical analysis of the distribution of pond molluscs in Western Canada. *American Naturalist* **70**: 237–44.

Noy-Meir, I. (1973). Data transformations in ecological ordination. I: Some advantages of non-centering. *Journal of Ecology* **61**: 329–41.

Ochiai, A. (1957). Zoogeographical studies on the soleoid fishes found in Japan and its neighbouring regions. *Bulletin of the Japanese Society of Scientific Fisheries* **22**: 526–30.

Odum, E. P. (1950). Bird populations of the Highlands (North Carolina) Plateau in relation to plant succession and avian invasion. *Ecology* **31**: 587–605.

Orloci, L. (1978). *Multivariate Analysis in Vegetation Research* (2nd edn). Junk, The Hague.

Patil, G. P. and Taillie, C. (1982). Diversity as a concept and its measurement. *Journal of the American Statistical Association* **77**: 548–67.

Pearson, T. H. (1975). The benthic ecology of Loch Linnhe and Loch Eil, a sea-lock system on the west coast of Scotland. IV: Changes in the benthic fauna attributable to organic enrichment. *Journal of Experimental Marine Biological Ecology* **20**: 1–41.

Persson, S. (1981). Ecological indicator values as an aid in the interpretation of ordination diagrams. *Journal of Ecology* **69**: 71–84.

Pielou, E. C. (1961). Segregation and symmetry in two-species populations as studied by nearest neighbour relations. *Journal of Ecology* **49**: 255–69.

Pielou, E. C. (1984). *The Interpretation of Ecological Data*. Wiley, New York.

Sanders, H. L. (1968). Marine benthic diversity: A comparative study. *American Naturalist* **102**: 243–82.

Schiffman, S. S., Reynolds, M. L. and Young, F. W. (1981). *Introduction to Multidimensional Scaling. Theory, Methods and Applications*. Academic Press, New York.

Seal, H. L. (1966). *Multivariate Statistical Analysis for Biologists*. Wiley, New York.

Sibson, R. (1979). Studies in the robustness of multidimensional scaling: Perturbational analysis of classical scaling. *Journal of the Royal Statistical Society*, Series B, **41**: 217–29.

Smith, W. and Grassle, J. F. (1977). Sampling properties of a family of diversity measures. *Biometrics* **33**: 283–92.

Sneath, P. H. A. and Sokal, R. R. (1973). *Numerical Taxonomy*. W. H. Freeman & Co., San Francisco.

Solomon, D. L. (1979). A comparative approach to species diversity. In *Ecological Diversity in Theory and Practice* (eds J. F. Grassle, G. P. Patil, W. K. Smith and C. Taillie), pp. 29–36. International Cooperative Publishing House, Fairland, Md.

Sørensen, T. (1948). A method of establishing groups of equal amplitude in plant sociology based on similarity in species content. *Biol. Skr. K. danske Vidensk. Selsk.* **5**(4): 1–34.

Taylor, L. R. (1986). Synoptic dynamics, migration and the Rothamsted Insect Survey. *Journal of Animal Ecology* **55**: 1–38.

Taylor, L. R., French, R. A., Woiwod, I. P., Dupuch, M. J. and Nicklen, J. (1981). Synoptic monitoring for migrant insect pests in Great Britain and Western Europe. I: Establishing expected values for species content, population stability and phenology of aphids and moths. In *Rothamsted Experimental Station, Annual Report for 1980*, Part 2, pp. 41–104. Harpenden, UK.

ten Berge, J. M. F. (1977). Orthogonal Procrustes rotation for two or more matrices. *Psychometrika* **42**: 267–76.

ter Braak, C. J. F. (1983). Principal component biplots and alpha and beta diversity. *Ecology* **64**: 454–62.

Tietjen, J. H. (1971). Ecology and distribution of deep-sea meiobenthos off North Carolina. *Deep-Sea Research* **18**: 941–57.

Trass, H. and Malmer, N. (1973). North European approaches to classification. In *Ordination and Classification of Communities* (ed. R. H. Whittaker), pp. 529–74. Junk, The Hague.

Tufte, E. R. (1983). *The Visual Display of Quantitative Information*. Graphics Press, Cheshire, Conn.

Tukey, J. W. (1977). *Exploratory Data Analysis*. Addison-Wesley, Reading, Mass.

Tukey, P. A. and Tukey, J. W. (1981). Graphical display of data sets in 3 or more dimensions. In *Interpreting Multivariate Data* (ed. V. Barnett), pp. 189–275. Wiley, Chichester.

Usher, M. B. (1981). Modelling ecological succession, with particular reference to Markovian models. *Vegetatio* **46**: 11–18.

Westhoff, V. and Maarel, E. van der (1973). The Braun–Blanquet approach. In *Ordination and Classification of Communities* (ed. R. H. Whittaker), pp. 617–726. Junk, The Hague.

Whittaker, R. H. (ed.) (1978a). *Classification of Plant Communities*. Junk, The Hague.

Whittaker, R. H. (ed.) (1978b). *Ordination of Plant Communities*. Junk, The Hague.

Williams, C. B. (1964). *Patterns in the Balance of Nature*. Academic Press, London.

Williams, E. D. (1978). *Botanical Composition of the Park Grass Plots at Rothamsted 1856–1976*. Rothamsted Experimental Station, Harpenden.

Williams, W. T. (ed.) (1976), *Pattern Analysis in Agricultural Science*. Elsevier, New York.

Williams, W. T. and Lambert, J. M. (1959). Multivariate methods in plant ecology. I: Association analysis in plant communities. *Journal of Ecology* **47**: 83–101.

Williamson, M. H. (1978). The ordination of incidence data. *Journal of Ecology* **66**: 911–20.

Wilson, M. V. (1981). A statistical test of the accuracy and consistency of ordinations. *Ecology* **62**: 8–12.

Woiwod, I. P. and Tatchell, G. M. (1984). Computer mapping of aphid abundance. In *Proceedings of the 1984 British Crop Protection Conference – Pests and Diseases*, pp. 675–83.

Wright, A. J. (1971). The analysis and prediction of some two factor interactions in grass breeding. *Journal of Agricultural Science* **76**: 301–6.

Yarranton, G. A. (1966). A plotless method of sampling vegetation. *Journal of Ecology* **54**: 229–37.

Appendix Matrix algebra

Any discussion of multivariate methods needs some notational basis to describe the various numerical operations. These operations range from the extremely complex, such as the singular value decomposition (Section A.8), to the relatively simple, such as the calculation of means. Fortunately, the ordinary user of multivariate methods need not be concerned with the details of the numerical calculations since these will nearly always be computerized; however, some form of notation is still required to understand the relationship between the many different methods. To avoid unnecessary complexity we use the concise notation of matrix algebra. While this may cause initial problems of understanding to some readers, it will soon be recognized that the notation has particular advantages for multivariate methods where the basic observational unit is a two-way table (or matrix) of, say, species by sites. Furthermore, many of the operations of matrix algebra have geometric interpretations, so this algebra is particularly appropriate for multivariate analysis where interpretation is often based on the geometry of a set of points in many dimensions.

In this Appendix we restrict ourselves to describing those areas of matrix algebra that are used to develop the multivariate methods discussed in this book. For a more comprehensive coverage the reader is referred to books by Graybill (1969) or Green and Carroll (1976).

A.1 MATRICES AND VECTORS

Matrices are rectangular arrays of real numbers arranged in rows and columns and are represented by bold upper-case letters, e.g.

$$\mathbf{X} = \begin{bmatrix} 5 & 3 & 0 \\ 1 & 9 & 2 \\ 5 & 2 & 4 \end{bmatrix}$$

The numbers of rows and columns of a matrix are important attributes and are often used to qualify the matrix: if **X** has n rows and p columns, **X** has size $(n \times p)$ and we write $\mathbf{X}\ (n \times p)$.

Vectors, denoted by bold lower case letters, are matrices with only one row or only one column; thus

$$\mathbf{v} = [4 \quad 5 \quad 22 \quad -3]$$

is a row vector of four elements, while

$$\mathbf{v} = \begin{bmatrix} 4 \\ 5 \\ 22 \\ -3 \end{bmatrix}$$

is the corresponding column vector. In this book we adopt the convention of regarding all vectors as column vectors unless otherwise qualified.

A single number is called a *scalar* and is denoted by a lower case letter, e.g. k. Scalars may be considered as special cases of matrices, with only one row and one column, or of vectors with only one element.

To refer to individual *elements* of matrices or vectors we use small letters with subscripts. Thus the element in the ith row and jth column of the matrix $\mathbf{X}$ $(n \times p)$ is denoted by the scalar x_{ij}, where the subscripts i and j lie between 1 and n, and 1 and p, respectively. Elements of vectors are denoted by a single subscript, i.e. v_i is the ith element of $\mathbf{v}$.

In some situations it is convenient to write a matrix as a rectangular array of submatrices, e.g.

$$\mathbf{X} = \begin{bmatrix} \mathbf{X}_{11} & \mathbf{X}_{12} & \mathbf{X}_{13} \\ \mathbf{X}_{21} & \mathbf{X}_{22} & \mathbf{X}_{23} \end{bmatrix}$$

and $\mathbf{X}$ is now considered as a *partitioned matrix*. Any partitioning must apply right across and right down the matrix $\mathbf{X}$, e.g. $\mathbf{X}_{12}$ and $\mathbf{X}_{22}$ must have the same number of columns. Then, if $\mathbf{X}_{ij}$ is of size $(n_i \times p_j)$, $\mathbf{X}$ is $(n \times p)$ where $n = n_1 + n_2$ and $p = p_1 + p_2 + p_3$. Partitioning is useful when, for example, the rows of the matrix can be grouped into a number of classes.

There are three commonly used geometric interpretations of vectors. Suppose that the vector $\mathbf{v}$ has p elements $(v_1, v_2, \ldots, v_p)$: then $\mathbf{v}$ can be interpreted as the location of p points along a single coordinate axis, viz. a straight line. Alternatively, $\mathbf{v}$ may be considered to give the location of a single point referred to p coordinate axes; when $p=2$ we have $\mathbf{v}=(v_1, v_2)$, the coordinates of a point in two dimensions which can be plotted on a piece of graph paper. When p is larger than three it becomes difficult to visualize such points; however, in order to understand the geometry it is usually sufficient to consider p less than or equal to three, and then generalize the ideas to higher dimensions. We will always make the assumption that the coordinate axes are mutually orthogonal, i.e. all at right angles to each other. Finally, as a variant of the latter interpretation, we can consider a vector as defining the direction, in p-dimensional space, of a line from the origin, which has coordinates $(0, 0, \ldots, 0)$, to the point $(v_1, v_2, \ldots, v_p)$. Such a vector has a length as well as a direction. From Pythagoras' theorem, the length is $r = \sqrt{(v_1^2 + v_2^2 + \ldots + v_p^2)}$, while the cosine of the angle between the vector and the ith coordinate axis is v_i/r; these latter values are called *direction cosines*.

If we consider the matrix $\mathbf{X}$ $(n \times p)$ as a collection of n row vectors, each of length p, we can interpret the elements of **X** as the coordinates of n points in p dimensions, x_{ij} being the jth coordinate of the ith point. Here the rows of **X** correspond to the points, and its columns correspond to the dimensions; thus if **X** is (10×2), its elements can be considered as the coordinates of 10 points that can be plotted on a piece of graph paper. The alternative interpretation, that **X** $(n \times p)$ gives the coordinates of p points in n dimensions, is rarely used.

A.2 PARTICULAR FORMS OF MATRICES

A *square matrix* has the same number of rows and columns, e.g. **X** $(n \times n)$ is a square matrix of order n. When a square matrix **X** is such that $x_{ij} = x_{ji}$ for all i and j, it is a *symmetric matrix*. A symmetric matrix **X** with $x_{ij} = 0$ for all $i \neq j$ is said to be a *diagonal matrix*, since its only non-zero elements occur on the leading diagonal. The diagonal matrix **D** of order n may be written as diag $(d_1, d_2, \ldots, d_n)$, or diag (**d**) where **d** is a vector of order n. A square matrix **X** with $x_{ij} = 0$ for all $i < j$ is known as a *lower triangular matrix* since its only non-zero elements occur on or below the leading diagonal. An upper triangular matrix is similarly defined to have all its non-zero elements on or above the leading diagonal. **A**, **B** and **C** below are examples of 4×4 symmetric, diagonal and lower triangular matrices respectively:

$$\mathbf{A} = \begin{bmatrix} 5 & 0 & 2 & 4 \\ 0 & 8 & 6 & 9 \\ 2 & 6 & -6 & -2 \\ 4 & 9 & -2 & 1 \end{bmatrix}, \quad \mathbf{B} = \begin{bmatrix} 6 & 0 & 0 & 0 \\ 0 & -8 & 0 & 0 \\ 0 & 0 & 3 & 0 \\ 0 & 0 & 0 & 10 \end{bmatrix}, \quad \mathbf{C} = \begin{bmatrix} 8 & 0 & 0 & 0 \\ 4 & 2 & 0 & 0 \\ 0 & 4 & 6 & 0 \\ 1 & 2 & 9 & 1 \end{bmatrix}$$

Note that, for a symmetric matrix, it is only necessary to specify the elements on and below the diagonal: we have adopted this convention in this book, so that **A** would appear as

$$\begin{bmatrix} 5 & & & \\ 0 & 8 & & \\ 2 & 6 & -6 & \\ 4 & 9 & -2 & 1 \end{bmatrix}$$

A matrix is of *block-diagonal* form if it may be partitioned as a square array of submatrices, where all submatrices lying off the diagonal are zero. Thus,

$$\mathbf{X} = \left[\begin{array}{cc|c|ccc} 5 & 3 & 0 & 0 & 0 & 0 \\ 4 & 9 & 0 & 0 & 0 & 0 \\ \hline 0 & 0 & 2 & 0 & 0 & 0 \\ \hline 0 & 0 & 0 & 8 & 0 & 1 \\ 0 & 0 & 0 & 7 & 2 & 3 \\ 0 & 0 & 0 & 5 & 4 & 9 \end{array}\right]$$

is block diagonal.

A.3 SIMPLE MATRIX OPERATIONS

Many of the natural operations and properties of ordinary scalar algebra do not transfer immediately to matrix algebra – some do not transfer at all. Furthermore, many matrix operations are only defined when the matrices are of a certain type; they are then said to be *conformable* for the particular operation, e.g. matrix addition.

Operations involving vectors are handled by treating the vectors as matrices with only one row, or only one column.

Matrix addition, as in $\mathbf{C}=\mathbf{A}+\mathbf{B}$, is only valid (conformable) when $\mathbf{A}$ and $\mathbf{B}$ have the same number of rows and the same number of columns. If $\mathbf{A}$ and $\mathbf{B}$ have dimension $(m \times n)$ the resultant $\mathbf{C}$ is also $(m \times n)$ and has elements given by $c_{ij}=a_{ij}+b_{ij}$. Note that, as in scalar algebra, the operation is commutative, i.e. $\mathbf{A}+\mathbf{B}=\mathbf{B}+\mathbf{A}$.

Matrix subtraction is analogous to matrix addition; when $\mathbf{A}$ and $\mathbf{B}$ are both $(m \times n)$ we have $\mathbf{C}$ $(m \times n)=\mathbf{A}-\mathbf{B}$ with elements $c_{ij}=a_{ij}-b_{ij}$.

Matrix multiplication is quite distinct from scalar multiplication. However, multiplying a matrix by a scalar is straightforward: the product $k\mathbf{A}$ has the same number of rows and columns as $\mathbf{A}$ and has its (i, j)th element equal to ka_{ij}. However, the product of two matrices, $\mathbf{C}=\mathbf{AB}$, is only defined when the number of columns of the first matrix equals the number of rows of the second. Thus if $\mathbf{A}$ is $(m \times n)$, $\mathbf{B}$ must have n rows, and if $\mathbf{B}$ is $(n \times p)$ the resultant $\mathbf{C}$ will be $(m \times p)$. To form c_{ij} we take the ith row of $\mathbf{A}$ and jth column of $\mathbf{B}$, and noting that both of these vectors will have n values, the corresponding elements of the vectors are multiplied and the sum of these products gives c_{ij}, i.e. $c_{ij}=\sum_{k=1}^{n} a_{ik}b_{kj}$.

For example, if $\mathbf{A}=\begin{bmatrix} 2 & 8 \\ 1 & -2 \end{bmatrix}$ and $\mathbf{B}=\begin{bmatrix} 6 & -2 & 0 \\ 1 & 3 & 4 \end{bmatrix}$, then

$$\mathbf{C}=\mathbf{AB}=\begin{bmatrix} 12+8 & -4+24 & 0+32 \\ 6-2 & -2-6 & 0-8 \end{bmatrix}=\begin{bmatrix} 20 & 20 & 32 \\ 4 & -8 & -8 \end{bmatrix}$$

By considering the relative sizes of $\mathbf{C}=\mathbf{AB}$ and $\mathbf{D}=\mathbf{BA}$, it is clear that $\mathbf{C}$ and $\mathbf{D}$ will not have the same size unless $\mathbf{A}$ and $\mathbf{B}$ are both square; for example, the product of a row vector and column vector is a scalar, while the product of a column vector is a matrix. Thus

$$[5 \quad -3 \quad 1]\begin{bmatrix} 2 \\ 4 \\ 1 \end{bmatrix}=10-12+1=-1$$

while

$$\begin{bmatrix} 2 \\ 4 \\ 1 \end{bmatrix} [5 \quad -3 \quad 1] = \begin{bmatrix} 10 & -6 & 2 \\ 20 & -12 & 4 \\ 5 & -3 & 1 \end{bmatrix}.$$

It is clear that matrix multiplication is not commutative so that **AB** will not generally equal **BA**; this is contrary to the property in ordinary (scalar) algebra where $ab = ba$ for all a, b. Indeed, this property does not hold, in general, even for symmetric matrices.

These first few matrix operations all have analogues in scalar algebra. One matrix operation that has no such analogue is *matrix transposition*, written as $\mathbf{A}'$ (or in some other texts as $\mathbf{A}^T$). If $\mathbf{A}$ is $(m \times n)$, its transpose, $\mathbf{B} = \mathbf{A}'$, is $(n \times m)$ and is defined as $b_{ij} = a_{ji}$. Clearly, transposition of a symmetric or diagonal matrix leaves it unchanged; transposition of a column vector changes it to a row vector, and vice versa.

Furthermore, if $\mathbf{C} = \mathbf{A} + \mathbf{B}$ then $\mathbf{C}' = \mathbf{A}' + \mathbf{B}'$; also if $\mathbf{B} = k\mathbf{A}$ then $\mathbf{B}' = k\mathbf{A}'$. By considering the formation of the matrix product $\mathbf{C} = \mathbf{AB}$, it can be seen that $\mathbf{C}' = (\mathbf{AB})' = \mathbf{B}'\mathbf{A}'$.

Operations involving partitioned matrices are defined in an obvious way. The partitioning must be conformable for the operation: for example, if we have the two matrices

$$\mathbf{X} = [\mathbf{X}_1 \mathbf{X}_2], \quad \mathbf{Y} = \begin{bmatrix} \mathbf{Y}_1 \\ \mathbf{Y}_2 \end{bmatrix},$$

where $\mathbf{X}$ $(m \times n)$ is partitioned into $\mathbf{X}_1$ $(m \times n_1)$ and $\mathbf{X}_2$ $(m \times n_2)$, then to form $\mathbf{Z} = \mathbf{XY}$, $\mathbf{Y}$ must have n rows and $\mathbf{Y}_i$, n_i rows. If $\mathbf{Y}$ has p columns, $\mathbf{Z}$ will be given by

$$\mathbf{Z}\ (m \times p) = \mathbf{X}_1\mathbf{Y}_1 + \mathbf{X}_2\mathbf{Y}_2.$$

A.4 SIMPLE GEOMETRY AND SOME SPECIAL MATRICES

In a number of multivariate methods it is usual to regard a data matrix $\mathbf{X}$ $(n \times p)$ as defining the coordinates of n points in p dimensions, i.e. the columns of $\mathbf{X}$ relate to p orthogonal coordinate axes. The location of the average point, or *centroid is* $\bar{\mathbf{x}}$ $(1 \times p)$ where

$$\bar{x}_i = \frac{1}{n} \sum_{k=1}^{n} x_{ik}.$$

The column vector which has n elements all equal to one is usually denoted by $\mathbf{1}_n$; so we have $\bar{\mathbf{x}} = 1/n\ \mathbf{1}'\mathbf{X}$ (we omit the subscript n, since the vector must have n elements for the multiplication to be defined).

For many methods it is convenient to move the points so that their positions relative to each other remain the same but their new centroid is at the origin.

This is achieved by subtracting the coordinates of the old centroid $\bar{\mathbf{x}}$ from the coordinates of each point. Thus if the new positions are in the matrix $\mathbf{Y}$ $(n \times p)$ we have $\mathbf{Y}=\mathbf{X}-\bar{\mathbf{X}}$ where every row of $\bar{\mathbf{X}}$ equals $\bar{\mathbf{x}}$, i.e. $\bar{\mathbf{X}}=\mathbf{1}_n\bar{\mathbf{x}}'$. Reconstructing the above steps we have

$$\mathbf{Y}=\mathbf{X}-\frac{1}{n}\mathbf{1}\mathbf{1}'\mathbf{X}=(\mathbf{I}-\mathbf{N})\mathbf{X}=\mathbf{H}\mathbf{X}.$$

The new matrices introduced are all of order n and defined as follows. The matrix $\mathbf{I}$ is the diagonal matrix with all of its diagonal entries being one; it has the property that the product of any matrix $\mathbf{X}$ by $\mathbf{I}$ leaves $\mathbf{X}$ unchanged, $\mathbf{X}=\mathbf{XI}=\mathbf{IX}$, and is thus analogous to unity in scalar algebra; it is called the *unit* or *identity matrix*. The matrix $\mathbf{N}$ is simply $1/n\ \mathbf{1}\mathbf{1}'$ and is the square matrix all of whose values are $1/n$. The matrix $\mathbf{H}$ is $\mathbf{I}-\mathbf{N}$ and is called the *centring matrix*; it has the property that premultiplication of $\mathbf{X}$ by $\mathbf{H}$ centres the points with coordinates in $\mathbf{X}$ to have their centroid at the origin.

A.5 MATRIX INVERSION

In scalar algebra the idea of dividing one number by another is taken for granted; thus the expression $c=a/b$, or its less commonly used form $c=ab^{-1}$, is well understood. It is often forgotten that a/b is only defined when b does not equal zero, i.e. when b^{-1} exists; however, this point is pertinent since it has an important analogue in matrix algebra.

There is no concept of matrix division: instead one multiplies one matrix by the inverse of another, e.g. $\mathbf{C}=\mathbf{AB}^{-1}$. Just as $bb^{-1}=1$ and $b^{-1}b=1$, the inverse of the matrix $\mathbf{B}$ is the matrix $\mathbf{B}^{-1}$ that satisfies $\mathbf{BB}^{-1}=\mathbf{B}^{-1}\mathbf{B}=\mathbf{I}$, where $\mathbf{I}$ is the identity matrix. Remembering the earlier discussion of matrix multiplication (Section A.3), it is clear that we can only have $\mathbf{BB}^{-1}=\mathbf{B}^{-1}\mathbf{B}$ when $\mathbf{B}$ is square; even then an inverse of $\mathbf{B}$ may not exist, just as the reciprocal of a scalar k does not exist when $k=0$. Matrices that do not have an inverse are said to be *singular*; conversely, invertible matrices are *non-singular*. We will return to this point in the next section.

The computation of matrix inverses is usually performed by a computer algorithm; for our purposes it is not necessary to know how this is done. However, a few results concerning matrix inverses are useful. The inverse of the diagonal matrix $\mathbf{D}=\text{diag}(d_1, d_2, \ldots, d_n)$ is given by $\mathbf{D}^{-1}=\text{diag}(d_1^{-1}, d_2^{-1}, \ldots, d_n^{-1})$; i.e. we invert each diagonal value, provided that none of these is zero, in which case the matrix $\mathbf{D}$ would be singular. The inverse of $k\mathbf{A}$ is $k^{-1}\mathbf{A}^{-1}$ provided that $k\neq 0$ and $\mathbf{A}$ is non-singular. Less straightforwardly, the inverse of a matrix product is given by $(\mathbf{AB})^{-1}=\mathbf{B}^{-1}\mathbf{A}^{-1}$.

A.6 SCALAR FUNCTIONS OF MATRICES

In this section we define three scalar values that can be associated with matrices; the first two of these exist only for square matrices, the third applies to any matrix.

The *trace* of a square matrix **A** is the sum of the diagonal elements of **A** and is written as trace (**A**) or tr(**A**). Thus $\text{tr}(\mathbf{A}) = \Sigma a_{ii}$.

Earlier we used the centring matrix **H** to obtain the matrix **Y** ($n \times p$) with centroid at the origin; let us now examine tr(**YY**′) in a little more detail. If we let $\mathbf{S} = \mathbf{YY}'$, then

$$t = \text{tr}(\mathbf{S}) = \sum_{i=1}^{n} s_{ii} = \sum_{i=1}^{n} \sum_{j=1}^{p} y_{ij}^2,$$

so that tr(**YY**′) is simply the sum of the squares of all elements of the matrix **Y**. If **Y** is considered as giving the coordinates of a set of n points in p dimensions, we can use Pythagoras' theorem to obtain the squared distance of the ith point from the origin as $d_i^2 = \sum_{j=1}^{p} y_{ij}^2$, which is simply the ith diagonal value of the matrix **S**. Thus t is the total squared distance of all the n points from their centroid (since, for **Y**, this is at the origin).

A few other results can be easily shown by considering the diagonal values of the resultants:

$$\text{tr}(\mathbf{A} + \mathbf{B}) = \text{tr}(\mathbf{A}) + \text{tr}(\mathbf{B})$$
$$\text{tr}(k\mathbf{A}) = k\ \text{tr}(\mathbf{A})$$
$$\text{tr}(\mathbf{AB}) = \text{tr}(\mathbf{BA}).$$

Particular cases of the last result are that $\text{tr}(\mathbf{YY}') = \text{tr}(\mathbf{Y}'\mathbf{Y})$ and that the sum of squares of the vector **v** is $\mathbf{v}'\mathbf{v} = \text{tr}(\mathbf{v}'\mathbf{v}) = \text{tr}(\mathbf{vv}')$.

The *determinant* of a square matrix is written as det(**A**) or $|\mathbf{A}|$. The determinant of a scalar is the value of the scalar; the determinant of **A** (2×2) is $a_{11}a_{22} - a_{12}a_{21}$. Determinants of higher-order matrices are usually given in terms of determinants of submatrices; thus when **A** is of order three,

$$|\mathbf{A}| = a_{11} \begin{vmatrix} a_{22} & a_{23} \\ a_{32} & a_{33} \end{vmatrix} - a_{12} \begin{vmatrix} a_{21} & a_{23} \\ a_{31} & a_{33} \end{vmatrix} + a_{13} \begin{vmatrix} a_{21} & a_{22} \\ a_{31} & a_{32} \end{vmatrix}.$$

As with matrix inverses, determinants are usually calculated by a computer using algorithms that need not concern us: the importance of determinants lies in their special uses and properties.

The determinant of a diagonal matrix is the product of its diagonal values. When **A** ($n \times n$) and **B** ($n \times n$) are square, we have: $|\mathbf{A}'| = |\mathbf{A}|$, $|k\mathbf{A}| = k^n|\mathbf{A}|$, and $|\mathbf{AB}| = |\mathbf{A}|\,|\mathbf{B}|$. This last property allows us to return to matrix inverses. **A** has an inverse only if a matrix $\mathbf{A}^{-1}$ exists such that $\mathbf{AA}^{-1} = \mathbf{I}$; this implies that $|\mathbf{A}|\,|\mathbf{A}^{-1}| = |\mathbf{I}|$ and since $|\mathbf{I}| = 1$, that $|\mathbf{A}^{-1}| = (|\mathbf{A}|)^{-1}$; thus the determinant of the inverse of **A** is the reciprocal of the determinant of **A**. When $|\mathbf{A}| = 0$, its reciprocal $(|\mathbf{A}|)^{-1}$ does not exist and it follows that the inverse $\mathbf{A}^{-1}$ does not exist, i.e. that **A** is singular. In fact the matrix **A** is non-singular if, and only if, $|\mathbf{A}| \neq 0$. With this useful result certain others follow immediately, e.g. the product of the two square matrices **A** and **B** is non-singular if and only if both **A** and **B** are non-singular.

The third scalar function is the *rank* of a matrix, written as rank(**X**); unlike the trace and determinant it is defined for all matrices, not just those that are square. Previously we have considered **X** ($n \times p$) as giving the coordinates of n points in p dimensions. For illustrative purposes let $p = 2$ and consider the two examples of Fig. A.1. Although the coordinates are for two dimensions the points, and the origin, may be in a subspace of fewer dimensions: the rank of a matrix is simply the dimensionality of this subspace. For example, in Fig. A.1(a) the points lie on a straight line through the origin and so have dimensionality one; however, in Fig. A.1(b) the points are in two dimensions, and the matrix is said to be of full rank. In Fig. A.1(c) the points lie on a straight

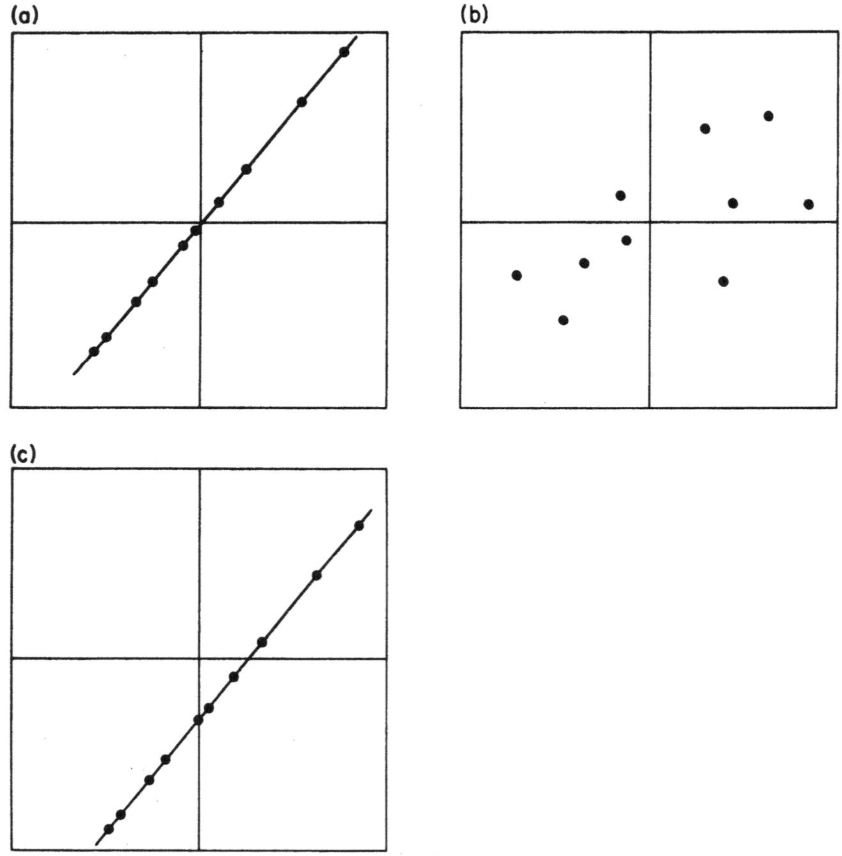

Fig. A.1 Scatters of points in two dimensions, from a data matrix **X** (10×2). In (a) the points and origin lie on a straight line: hence **X** has rank 1. In (b) the points are not linearly dependent: so **X** has rank 2. In (c) the line does not include the origin: so **X** has rank 2.

line; however, this line does not pass through the origin, so **X** has rank 2. If **X** were to be centred (using the centring matrix **H**), the resulting coordinates would be as in Fig. A.1(a), so that **HX** would have rank one.

An alternative, but equivalent, definition of rank is the number of linearly independent rows and columns of **X**. In Fig. A.1(a) the points lie on a line through the origin, so that $x_{i2} = cx_{i1}$: thus **X** has only one independent column (either the first, or the second). For **X** (2×3) we have two points in three dimensions and these points must lie on a plane through the origin, i.e. rank (**X**) is at most 2. From this argument, we can see that rank (**X**) $\leqslant$ min (n, p) and rank (**X**$'$) = rank (**X**). Another useful result is that rank (**XY**) $\leqslant$ min (rank (**X**), rank (**Y**)).

A.7 ORTHOGONAL MATRICES

Consider the situation where the n points in two dimensions, whose coordinates are given by **X** $(n \times 2)$, lie on a straight line through the origin, as in Fig. A.1(a). Now an adequate summary of the relative positions of the points would be the coordinates of the points on the straight line, obtained by rotating the original axes so that the new axis coincides with the straight line. Here all the variation in the data is described on this (first) new axis, making a second axis superfluous. However in general, when there is a certain amount of scatter about this line, the scatter would be exhibited in a second new axis at right angles (*orthogonal*) to the first. If the first new axis is at an angle θ to the first old axis, then $a = \cos\theta$ and $b = \cos(90 - \theta) = \sin\theta$ are the direction cosines of the first new axis to the old axes; likewise, the directions cosines of the second new axis to the old axis are $-b$ and a. The positions of the points relative to the new axes are now:

$$y_{i1} = ax_{i1} + bx_{i2},$$
$$y_{i2} = -bx_{i1} + ax_{i2}.$$

In matrix notation

$$\mathbf{Y} = \mathbf{XP}, \text{ where } \mathbf{P} = \begin{bmatrix} a & b \\ -b & a \end{bmatrix}.$$

The matrix **P** defines the (orthogonal) rotation of the axes, and is called an *orthogonal matrix*. Because $a^2 + b^2 = \sin^2\theta + \cos^2\theta = 1$, the sum of squares of each row and column of **P** is unity; also the sum of products of pairs of rows and pairs of columns is zero.

As would be expected, this generalizes to n dimensions, where n old axes are to be rotated to n new axes. Any matrix **P** $(n \times n)$ is orthogonal, and thus defines an orthogonal rotation, if its (i, j)th element gives the direction cosine of the jth new axis relative to the ith old axis. If we swop the terms 'new' and 'old', the orthogonal matrix defining the reverse rotation is simply **P**$'$. Applying the two rotations, first **P** then **P**$'$, clearly leaves the situation unchanged, $\mathbf{Y}=\mathbf{XPP}'=\mathbf{X}$; thus $\mathbf{PP}'=\mathbf{I}$, and so $\mathbf{P}'$ is the inverse of $\mathbf{P}$, and vice versa (hence $\mathbf{P}'\mathbf{P}=\mathbf{I}$). Since $|\mathbf{P}|=|\mathbf{P}'|$ and $|\mathbf{P}|\,|\mathbf{P}'|=|\mathbf{PP}'|=|\mathbf{I}|=1$ we have the useful result that $|\mathbf{P}|=\pm 1$. If we let the ith column of $\mathbf{P}$ be denoted by $\mathbf{p}_i$, it is clear from $\mathbf{P}'\mathbf{P}=\mathbf{I}$ that $\mathbf{p}_i'\mathbf{p}_i=1$ and $\mathbf{p}_i'\mathbf{p}_j=0$ for $i\neq j$, as for the 2×2 example above. This condition among the set of vectors $\mathbf{p}_i$ is called *orthonormality*: the prefix 'ortho-' arises from their orthogonality, i.e. $\mathbf{p}_i'\mathbf{p}_j=0$; the term '-normality' indicates that they are *normalized* to have unit sum of squares, i.e. $\mathbf{p}_i'\mathbf{p}_i=1$.

A.8 MATRIX DECOMPOSITIONS

Many of the multivariate methods described in this book use one of two related matrix decompositions. By a decomposition we mean the expression of a matrix as the product of two or more matrices, these latter matrices being of particular types.

The *singular value decomposition* (SVD) holds for any matrix $\mathbf{X}\,(n\times p)$, where we assume $n\geqslant p$ (when $n<p$, similar results apply to the decomposition of $\mathbf{X}'$). The SVD is unique and is given by

$$\mathbf{X}=\mathbf{USV}',$$

where $\mathbf{U}$ is of the same size as $\mathbf{X}\,(n\times p)$, and $\mathbf{S}$ and $\mathbf{V}$ are square matrices of order p. The matrices have the following properties: the columns of $\mathbf{U}$ are orthonormal, i.e. $\mathbf{U}'\mathbf{U}=\mathbf{I}$; the matrix $\mathbf{V}$ is orthogonal, i.e. $\mathbf{V}'\mathbf{V}=\mathbf{VV}'=\mathbf{I}$; the matrix $\mathbf{S}$ is diagonal and its diagonal values are all non-negative and arranged in descending order of magnitude, i.e. $s_1\geqslant s_2\geqslant\ldots\geqslant s_p\geqslant 0$. The values $(s_1,\ldots,s_p)$ are called the *singular values of* $\mathbf{X}$. By expanding the SVD, the elements of $\mathbf{X}$ may be expressed as

$$x_{ij}=u_{i1}s_1v_{j1}+u_{i2}s_2v_{j2}+\ldots+u_{ip}s_pv_{jp},$$

which is called a *multiplicative model*, since all the terms on the right-hand side are products of individual factors: because of the orthonormality of $\mathbf{U}$ and $\mathbf{V}$, these multiplicative terms are independent. If the first $r(<p)$ terms are used to form the matrix $\mathbf{X}(r)$ of rank r, the Eckart–Young theorem (Eckart and Young, 1936) shows that $\mathbf{X}(r)$ is the best (least-squares) approximation to $\mathbf{X}$ of rank r.

The *spectral decomposition* of a matrix applies only when $\mathbf{A}$ is symmetric (order n). Then the spectral decomposition is

$$\mathbf{A}=\boldsymbol{\Gamma}\boldsymbol{\Lambda}\boldsymbol{\Gamma}',$$

where $\boldsymbol{\Gamma}$ is an orthogonal matrix of order n, and $\boldsymbol{\Lambda}$ is a diagonal matrix with diagonal elements $\lambda_1\geqslant\lambda_2\geqslant\ldots\geqslant\lambda_n$. The scalars λ_i are called the *eigenvalues* of $\mathbf{A}$ (other terms often used are *latent roots* or *characteristic values*). Each column of $\boldsymbol{\Gamma}$ is an *eigenvector* (or *latent vector*) of $\mathbf{A}$.

This decomposition is particularly useful and allows us to tie together some seemingly unrelated topics. The trace of $\mathbf{A}$ is simply the sum of the eigenvalues:

$\text{tr } \mathbf{A} = \text{tr}(\boldsymbol{\Gamma\Lambda\Gamma'}) = \text{tr}(\boldsymbol{\Lambda\Gamma'\Gamma}) = \text{tr } \boldsymbol{\Lambda}$. Similarly, we find the determinant of **A** is the product of all the eigenvalues:

$$|\mathbf{A}| = |\boldsymbol{\Gamma\Lambda\Gamma'}| = |\boldsymbol{\Gamma}|\,|\boldsymbol{\Lambda}|\,|\boldsymbol{\Gamma'}| = |\boldsymbol{\Lambda}| = \prod \lambda_i$$

Clearly **A** is non-singular (and can be inverted) if and only if all its eigenvalues are non-zero. The number of non-zero eigenvalues gives the rank of **A**: by considering $\boldsymbol{\Gamma\Lambda\Gamma'}$ we see that each zero eigenvalue reduces by one the number of independent eigenvectors used to form **A**. If **A** is non-singular we can define its inverse as follows:

$$\mathbf{A}^{-1} = (\boldsymbol{\Gamma\Lambda\Gamma'})^{-1} = \boldsymbol{\Gamma}'^{-1}\boldsymbol{\Lambda}^{-1}\boldsymbol{\Gamma}^{-1} = \boldsymbol{\Gamma\Lambda}^{-1}\boldsymbol{\Gamma}',$$

i.e. the eigenvectors of $\mathbf{A}^{-1}$ are the same as those of **A** and the eigenvalues of $\mathbf{A}^{-1}$ are the reciprocals of the eigenvalues of **A**. Similarly, the matrix $\mathbf{A}^{-1/2}$ (the symmetric inverse square root of **A**) is given by $\mathbf{A}^{-1/2} = \boldsymbol{\Gamma\Lambda}^{-1/2}\boldsymbol{\Gamma}'$, where the ith element of $\boldsymbol{\Lambda}^{-1/2}$ is $1/\sqrt{\lambda_i}$.

The singular value and spectral decompositions are closely linked. From the SVD of **X** we can obtain the spectral decomposition of the symmetric matrix $\mathbf{X'X}$: if $\mathbf{X} = \mathbf{USV'}$ then $\mathbf{X'X} = (\mathbf{USV'})'(\mathbf{USV'}) = \mathbf{VS(U'U)SV'} = \mathbf{VS^2V'}$, since $\mathbf{U'U} = \mathbf{I}$, and **S**, and therefore $\mathbf{S}^2$, are diagonal. Comparing this with the spectral decomposition, $\mathbf{X'X} = \boldsymbol{\Gamma\Lambda\Gamma'}$, shows that the eigenvectors $\boldsymbol{\Gamma}$ of $\mathbf{X'X}$ are the same as the *right singular vectors* **V** of **X**; also the eigenvalues $\boldsymbol{\Lambda}$ of $\mathbf{X'X}$ are the squares of the singular values **S** of **X**. This latter equivalence implies that none of the eigenvalues of $\mathbf{X'X}$ may be negative: such matrices are said to be *positive semi-definite*.

A modified form of the SVD of **X** allows us to show a parallel result for $\mathbf{XX'}$. Writing $\mathbf{X} = \mathbf{USV'}$, where $\mathbf{U}$ $(n \times n)$ is now orthogonal, as is $\mathbf{V}$ $(p \times p)$, and $\mathbf{S}$ $(n \times p)$ is such that the leading diagonal values s_{ii} contain the singular values of **X**. Proceeding as above,

$$\mathbf{XX'} = \mathbf{USV'VS'U'} = \mathbf{U\Lambda^*U'},$$

where $\boldsymbol{\Lambda}^*$ $(n \times n)$ is diagonal and $\lambda_i^* = s_{ii}^2$. Then the (non-zero) eigenvalues of $\mathbf{XX'}$ are the same as those of $\mathbf{X'X}$; also the eigenvectors of $\mathbf{XX'}$ are the full set of *left singular vectors* of **X**. The first p columns of the full matrix $\mathbf{U}$ $(n \times n)$ are those usually obtained as the left singular vectors from the SVD of **X**.

A.9 CONCLUSION

We cannot stress too heavily the benefits of a sound understanding of matrix algebra. As well as providing a notational basis for multivariate analysis, the theory can be used to express concisely various simple operations, such as the calculation of group means. A computer program, such as Genstat, which allows an easy specification of matrix operations, provides the ecologist with a very powerful tool for a wide variety of data manipulation and analysis.

Index

MIX
Papier aus verantwortungsvollen Quellen
Paper from responsible sources
FSC® C105338
FSC
www.fsc.org

If you have any concerns about our products,
you can contact us on
ProductSafety@springernature.com

In case Publisher is established outside the EU,
the EU authorized representative is:
Springer Nature Customer Service Center GmbH
Europaplatz 3, 69115 Heidelberg, Germany

Printed by Libri Plureos GmbH
in Hamburg, Germany